NUCLEAR CULTURE

NUCLEAR CULTURE

Living and Working in the World's Largest Atomic Complex

PAUL LOEB

Coward, McCann & Geoghegan, Inc.
New York

 Published on the same day in Canada by General Publishing Co. Limited, Toronto.

The author gratefully acknowledges permission to quote from:

Jefferson Airplane, "Volunteers," from their album "Volunteers" and from the "Woodstock" album. Copyright © 1969 Icebag Music Corp. Reprinted by permission of the publisher. All rights reserved.

"Ramblin' Rose." Words by Robert Hunter. Music by Jerry Garcia. Copyright © 1972. Ice Nine Pub. Co. Reprinted by permission. All rights reserved.

Library of Congress Cataloging in Publication Data

Loeb, Paul, Rogat, date.
Nuclear culture, living and working in the world's largest atomic complex

1. Pacific Northwest Laboratory. I. Title.
TK9024.W2L64 1982   306'.4   81-3194
ISBN 0-698-11104-4   AACR2

PRINTED IN THE UNITED STATES OF AMERICA

Contents

NUCLEAR CULTURE

PRELUDE

Atomic Soap: On the Job with the Young and the Restless

> "An alert careful operator is always important to assure our customer of high quality products."
> —training manual, Hanford plutonium separations plant

> "Look, if I don't wear a mask then you shouldn't either."
> —Hanford radiation monitor to a plutonium separations plant operator

Although the container fire was extinguished, Julie's mouth, nose, ears and hair still set every alpha particle detector screaming. Her white protective coveralls were totally contaminated. Keeping them on meant additional exposure risk each minute.

But instead of changing into clean clothing, Julie froze—shaking her head and repeating "I'm not undressing, I'm not undressing," while the radiation monitors and her fellow nuclear chemical operators alternately comforted and yelled at her. The accident wasn't her fault. She was embarrassed to weigh 190 pounds at age twenty-one. Why should she let the men see her in her underwear?

She stood terrified until someone found an old canvas cloth she could strip behind. She undressed while he held it up and the other men turned their backs. She went off with a radiation monitor to be decontaminated.

Both Julie and her co-worker Robert had followed all standard procedures in checking the triple-wrapped brass can containing plutonium and Uranium-235 out of the Special Nuclear Materials vault where it was safeguarded against theft or seizure by what the institution referred to as "hostile factions." They'd exercised proper care placing the can in the little red wagon with the special separation rack that insured its contents would not spontaneously "go critical" and begin a chain reaction. They'd put on their protective breather masks, just as they were supposed to. They'd transferred the can by means of a special port into a sealed glove box, and opened it just as usual, to check its contents, then switched them into a one-pound slip-lid tuna can. The new can was then weighed, taped and sealed, so the "product" it contained could be shipped from this nuclear reservation that sits, half the size of Rhode Island, in the eastern Washington desert, and delivered to the weapons and research labs of Los Alamos, New Mexico.

Only afterward—when the container was back outside the glove box, once more closed and once more wrapped in the customary two plastic outer bags (and when, since it was now safely sealed, Robert had removed his breather mask and Julie had begun to take off hers)—did it begin flaming like a Roman candle packed with carcinogenic dust.

The rule books explained coolly and rationally how operators should deal with situations like this by leaving the room, calling in an alarm, notifying their supervisor—and only then, if the fire was small and involved no radioactive materials, attempting to damp it. So perhaps it was poor judgment for Robert to use the chemical extinguisher, knock the can off the table and spray its contents all over the room. But he did put the fire out, and if the building's automatic CAM (Constant Air Monitoring) alarms were now going off in concert like a pack of howling monkeys it was the fault, not of the two operators, but of whomever had misplaced the documents indicating that the can contained an abnormally flammable mix of plutonium, uranium and kerosene solvent.

Julie and Robert scrubbed off in a decontamination room where—because the thirty-year-old building was originally designed for male workers only and the showers were side by side with no intervening barriers—Julie had to wait primly outside until Robert and a male radiation monitor, who was also contaminated, were finished. Then, because this fire involved personnel contamination, an ambulance brought the shielded bubbletop stretcher known as the Nuclear Accident Carrier. But the batteries powering the carrier's air supply turned out to be dead. A search of the ambulance turned up no replacements. Julie and Robert ended up riding in a regular government car to the "200 West" area first aid station.

At first Julie and Robert resisted the special shots that would allow them to pass out through their urine most of the plutonium they'd inhaled, and thus ingested into their bloodstream. When Robert at last submitted, the nurse who came to the decontamination room kept missing his veins until she finally gave up in near panic. But after they at last received the treatment, the doctors said they'd be fine.

Despite its happy ending, the incident upset a twenty-nine-year-old co-worker of Julie and Robert's named Amy Downing. She, like them, was used to going along day by day, following the rules as best she could, assuming and hoping everything would work out for the best. She'd never really viewed her work as involving potential danger. But dealing with burning plutonium wasn't on her agenda each morning. And since the can could just as easily have flared up in her face, she wondered whether particles had gotten into the two operators' lungs—and if so what damage had been done. She thought perhaps everyone in Z plant—the plutonium processing facility where they worked—should worry a bit more than they had.

There was also more to be apprehensive about than physical harm, and it was for this reason Amy and all the other technicians and operators praised Julie and Robert's actions to all the supervisors and investigators. Rules violations that could cause harm were "contacts." Three contacts, or one if it was serious enough, and you'd find yourself gone from your $17,000, $20,000 or $23,000 a year job here. You'd be back as an unskilled high school graduate making $4.00 an hour at McDonald's, the local french fry plant, or some factory like the one in Seattle Amy had worked at—opening and shutting a furnace door all day to pop plastic gears in and out, logging sixty hours a week just to keep the kids in jeans and hamburger.

For the next dozen weeks, Amy and the other Z plant employees worked rotating shifts decontaminating the walls, ceilings and floors of the room and adjacent hallway that the accident had "crapped up" (the universal Hanford term referring to the contamination of tools, people or physical environments). Had there been an Atomic Maid Service to match the Atomic Bowling Lanes, Atomic Body Shop and Atomic TV Repair in the neighboring town of Richland, they might have considered calling them in. Though perhaps some dreamer in the 1950s envisioned a George Jetson-styled flying automatic vacuum cleaner, that item had remained unrealized. The operators—wearing their filtered breather masks for protection—ended up scrubbing away the contamination with soap, rags and elbow grease just like any not-so-happy housewife in any soiled suburban kitchen.

When Julie and Robert returned to work a few days after the fire, they were greeted with the predictable remark "If you think I'm going to shake hands with you, you're crazy," then relegated for an indeterminate period to a supply room far from all radioactive zones. As their plutonium bond left them more edgy than comradely, Julie was soon telling everyone that Robert was "off his rocker," and he started teasing her about her weight.

Other times Julie was stronger than any of them—when the young supervisors, stuffed like Thanksgiving turkeys with their self-righteous book learning, began complaining unfairly about her performance or that of her fellow workers, and Julie stood up to say, "We're not wrong, you're wrong. This is the way we always do it." Of course sometimes the supervisors had a point. Maybe people should understand the rules better and the reasons behind them, instead of always going along with whatever the old-timers said was correct. But young Ph.D.'s, however much they jumped at other jobs here at the U.S. Department of Energy's Hanford site, seemed to shun ongoing work in this heart of the hands-on plutonium world—and that was the fault of no one here. Maybe, Amy thought, Julie had been scared in those first moments following the accident, of some nameless future damage either to her body or to the kids she hoped someday to have. Maybe Amy's own current boyfriend (a local teacher) was right in suggesting that even the regular allowable radiation doses might be dangerous—and she wished she remembered her latest monthly count and what it meant. But her two aunts had died of cancer long before atomic plants were even invented. Her father had worked here his entire adult life, yet lived to raise six kids and look forward to a retirement spent hunting and fishing in perfectly good health. Her worrying was probably just part of the chain of migraines she'd been getting from spending too much time with her breather mask on. And as for Julie—well, Amy liked her too much to ask if the accident made her worry.

Robert, on the other hand, had always been an obnoxious young buck: racing his Toronado to work as if he were Andy Granatelli; bragging to the other guys about how much dope he smoked and (after his divorce last year) how much the numerous women he'd fucked had admired his style; goofing off when the higher-ups were gone and when they were around volunteering for any stupid job that would prove he was tough. Robert had lost his hot dog attitude now—and in fact seemed almost mopey—but Amy remembered when his carelessness had led to her only contact.

It was right after Three Mile Island and a general tightening of procedures was making all the Z plant workers jumpy. She and Robert were transferring buttons of plutonium oxide from the tuna cans into plastic bags—locking each inside almost as you would a hamburger into a freezer-ready Seal-A-Meal. Since operators couldn't do the work in the open, the nuclear industry designed special "hoods" or "glove boxes" where they looked through glass windows, placed their hands inside thick, built-in rubber gloves and performed the required tasks without contamination danger. Although they were supposed to keep all the buttons ten inches apart in separate bags, the sealing took so long they could only make their daily quota by transferring five at once; which is what everyone did, just as they cut corners by not monitoring themselves each time they passed one of the mandatory checkpoints.

Things were going along as usual when Amy left the room to get some extra bags. Robert was supposed to stop work until she came back—security rules were quite strict about requiring two operators present. But he went ahead alone. One of the polyethylene bags, which he had failed to check for brittleness, developed a pinhole leak. Amy returned to find the CAM alarm going off, the room crapped up and a radiation monitor surveying for contamination. Though regulations for preventing contamination spread required Amy to stay out of any zone where an alarm was going off, it was her product batch and her shiftmate inside. She went in anyway to see what had happened.

Two weeks later Amy got a contact slip, but Robert didn't. Amy thought initially it was because she'd been dating the building superintendent and had broken it off. Then she decided she *had* disobeyed the rules, and admitted that with federal inspectors crawling all over the place, writing her up was probably good group control.

When Amy came to Hanford she expected to find young scientists like the ones in the TV ads: bloodless and efficient, white-coated and white in character and soul, displaying neither private idiosyncrasies nor private lusts. But most young guys here were nuclear operators who shared Robert's cocky, on-the-make and overpowered carelessness—like the union steward who told Amy if she didn't go out with him he'd dent her car or cite her for minor violations, or the part-time drug dealer whose entrepreneurial spirit brought the FBI to his house.

Amy didn't like the hustles, flirtations and sexual intrigues. There were, of course, other, older workers: the Hanford veterans, now mostly in their sixties and ready to retire with their travel trailers, boats, lakeside cabins and the old ladies to whom at least they'd remained largely faithful through all these years. Though many had no doubt been exposed to enough radiation to make an elephant glow, they went dourly about their day-to-day business, complaining about regulations this, regulations that, and about how if something got crapped up in the old days all you had to do was scrape it off in the desert sand outside. They seemed content in a manner few of her generation could be, and they were certain everything was fixable and nothing worth getting their blood up about—except perhaps the far-off Russkies or the all-too-present assaults on the nuclear industry by backpackers, whale watchers and Robert Redford types. But although Amy could learn from these men (and they made far better teachers than did the fresh-pressed, calculator-in-the-pocket new engineering graduates), and although she could work with them, they could not and did not make Z plant a place where she felt at home.

On bad days—when the inspectors were overinspecting, the men all hung over and the women all on Valium, when someone had walked accidentally through one of the security-sealed doors and brought down the Mod Squad (a super secret team of S.W.A.T. types who appeared within minutes wearing flak jackets and pointing machine guns at everyone)—on those days Amy saw nothing but an endless parade of traps. Sure, Millie the receptionist would always be there to furnish a cheery, optimistic horoscope. But Amy would come back from struggling with some mixing, percolation or separations process, would try rereading the training manuals and rule books to see where she'd gone wrong, and would end up nauseous from overload and confusion. "The destructive power of nuclear weapons is well known," explained this paper world. "Protection of the plutonium is very important . . . and should we lose control only history will tell the outcome." "If taken hostage be alert, maintain a good relationship with your captors, do not discuss what action may be taken by the Company . . . " What was this stuff about hostages, the Company and history? Amy wasn't the bionic woman, just an ordinary twenty-nine-year-old mom trying to do the best she could for herself and for her kids. She didn't want to learn a new kind of geometry to prevent plutonium from concentrating in the wrong, potentially critical configurations. If "inactive hoods containing greater than 100 grams fissile material require a 91 centimeter minimum spacing between the hoods and any fissile material near them" . . . then what did that have to do with the ten pounds Amy wanted to lose, with the payments on her brand new gold Capri, or with the aggravation from all the guys to whom she was either a hustling, overperforming chick or a stupid broad who couldn't get anything right? She'd put the book down at times like this, decide to blank out all hassles, go by what the old hands said and pray she could do her job well enough to not get contacted.

Those days and those depressions always passed, though. The work would go well, albeit a little boringly. She'd join her fellow operators in playing Hearts on swing-shift dinner breaks. They'd pitch in for gifts when some worker or worker's wife had a baby. And as if to cap and redeem the year, there was always the annual office Christmas party.

At the West Richland Elks Club, the old hands stepped out of their Buicks and Pontiacs wearing bolo ties, crew cuts and either cowboy boots or wing-tip oxfords. They swung and waltzed their wives on the dance floor in a manner that showed they'd had years and years of Saturday night stepping out. They talked of trout fishing and of grandchildren.

The young guys came in Camaros, four-wheel-drive pickups and even some Porsches. Some wore jeans, some wore suits. Gold chains rippled across their necks and chests. Their hair fell below their shoulders. If they weren't married to sweet ingenues just beginning to grow restless and uncertain from boredom, they drank at the bar and flirted across the room with women co-workers whose switch from coveralls to cocktail dresses had turned them desirably slinky and kittenish.

As the culminating event, a Santa Claus MC handed out a variety of joke prizes: "A pair of safety glasses to John who never wears his," "A glove because Rick burned his hands last year," and suntan lotion for Roy who was always standing right next to the radiation sources. Since the awards combined the best aspects of gossip, bragging recollection and childhood play, people laughed and clapped at each. Topping it all, Julie and Robert got T-shirts saying I'M HOT STUFF.

PART I

THE OLD HANDS

"They presented us with what they needed and we went out and built it . . ."—Hanford engineer

1
The Tinkerers

Z plant, where Amy and her colleagues work, is part of an atomic reservation which includes the world's first full-scale reactors and largest known radioactive waste storage site, a billion-dollar test facility for the breeder technologies that represent the fission industry's best future hope and three commercial nuclear plants still under construction—which if and when completed will be the most costly reactors in history.

From the air Hanford resembles a series of silver lunar cities surrounded by brown channeled desert. Located in the southeast corner of Washington State, the complex is bounded to the north and to the east by the giant U which the Columbia River makes in beginning the final stage of its journey from the Canadian Rockies to the Pacific. To the south, the Yakima River comes in to merge with the Columbia and to separate the town of Richland—immediately adjacent to the Hanford site—from Kennewick and Pasco, the other of the "Tri-Cities" where Hanford workers now live. To the west, Rattlesnake Mountain rises bare and spectral, hiding the desert flats and irrigated orchards of the Yakima Valley, then the Cascade Range, and—200 miles to the northwest—the city of Seattle. Mount St. Helens, assumed to be dormant when the reservation was founded, lies 120 miles to the west.

Hanford was still a tiny farm town when Hungarian physicist Leo Szilard drove out in the summer of 1939 to Albert Einstein's Long Island summer home. Szilard—together with two colleagues who accompanied him, Eugen Wigner and Edward Teller—worried that a nuclear bomb was possible, that the Germans were working on it and that the Allied nations needed a countervailing threat. He asked Einstein to use his prestige to convince Roosevelt of the need for an American atomic weapons program. When Einstein agreed, the letter he wrote began a process which—after two years of secret discussions between scientists, government officials and generals—led to the creation of the Manhattan Project.

Under the project's auspices, the world's first controlled chain reaction took place on December 2, 1942, in the Stagg Field test facility of Chicago's Metallurgical Laboratory. The laboratory's scientists had already begun developing methods to separate out the plutonium that would be created in this reaction. And because it appeared as if an atomic weapon could be created either from plutonium or from the scarce uranium isotope U-235, the project's directors decided to pursue both approaches simultaneously. They found an assembly site for the actual weapons at Los Alamos, New Mexico. Special facilities were to be built at the already existing research site of Oak Ridge, Tennessee, to separate out the U-235 from the far more common isotope U-238. Now the project needed a secure and geographically distanced location for the reactors that would produce the weapons-grade plutonium.

They found this isolation at Hanford, along with access to ample cooling water from the Columbia, to transcontinental rail transport from the Northern Pacific line that ran through the nearby town of Pasco and to nearly unlimited electrical power from the recently completed Grand Coulee Dam. When Army surveyors arrived in January 1943 to dig holes and test the bedrock for geological stability, local residents joked about forthcoming bonanzas from oil leases. But on February 23, a federal judge issued an expropriation order under the War Powers Act. A few days later curt notices arrived, giving the 1500 farmers—who had been cultivating irrigated orchards and vineyards in the valley surrounding the old towns of Hanford and White Bluffs—from fifteen to thirty days to leave their homes. Like the Japanese forced to relocate to Manzanar, they had no time to argue or resist. A few were permitted to return and complete their harvests, or, if they took Hanford jobs, to remain temporarily in houses on the edge of what was now officially the Hanford Engineering Works. A few defended their land briefly with shotguns. But the bulldozers, which the farmers referred to later as "them giant scoopmobiles," knocked down the houses and barns. Workers began immediate construction. Follow-up letters explained that even shrubs and trees were now government property and could not be removed.

Because the Manhattan Project did not have massive numbers of personnel directly under its command, it worked through the Army Corps of Engineers, and the engineers brought in the Du Pont corporation as the prime contractor to build and operate three plutonium production reactors and the accompanying facilities for separations, processing and fuel manufacture. Du Pont recruited the necessary workers from their existing, nonnuclear facilities in other states and from War Manpower Commission ads printed in newspapers and posted at government institutions throughout the country.

Since the Hanford project was top secret, the incoming men and women were told only that they were going to a nameless eastern Washington location where they would earn high wages, have living facilities provided and make an important contribution to the war effort. They came by train or car—one man had thirty-seven flat tires on the way up from Borger, Texas—and settled down in a newly built construction camp along with 45,000 other workers.

Like any construction boomtown anywhere, the Hanford camp had its brawling, gambling and drinking, so much so that Walter Winchell aroused the ire of the security people by suggesting in a column that mothers prohibit their daughters from coming here. Workers ate in huge mess halls, loading their plates from giant platters of pork chops and steaks and holding the platters up to be refilled each time they emptied. They took ten days to build a 4,000-person recreation hall, then attended movies in it and danced to top visiting performers like Kay Kyser, Tommy Dorsey and Benny Goodman. Except for families living in a 4,000-unit trailer camp, men and women slept in separate, racially segregated barracks, with fences surrounding those for the women. Because of this, local firemen blew their whistles each time they returned to their empty stations so as to chase out trysting couples, and there were rumors of midnight liaisons through fences. Nonworking wives had to remain back home or seek lodging in nearby farm towns.

But while most workers labored forty-eight hours a week building the reactors, the Corps of Engineers was also building a more permanent town for the production people. Expanding the 300-person village of Richland, 25 miles south of the Hanford camp, they brought in prefabs and threw together 4,000 clapboard military post houses—with letter designations indicating models of varying desirability that were allotted supposedly by family size, but in fact largely by status. They furnished each with identical chairs, tables, rugs and linens. They laid out streets which they named after military men (Panama Canal creator G. W. Goethals, his assistant Edgar Jadwin and imperial navy advocate Admiral Alfred Thayer Mahan) and after the newly planted elm, cottonwood and beech trees which the new residents hoped would flower and make this desert, jerry-built compound a home. Later—after the bomb was dropped and security eased—they added a court of streets named Proton, Argon and Nuclear lanes.

Because construction operations uprooted existing ground cover, dust storms were a constantly recurring discomfort. People lined up after each of these "termination winds" to resign their jobs and leave; so many terminated that Hanford went through over 140,000 employees in two years. The scarcity of support facilities created hour-long lines for haircuts, movies, cashing paychecks. Richland's houses were built so quickly that kids played by dropping marbles and watching them roll to the low point on the living room floor. And because the newly planted trees were still saplings, the atomic pioneers were known to hang Christmas ornaments on the ever-present tumbleweeds.

Despite all obstacles, B reactor became operational in September 1944, thirteen months after its construction began. It was joined shortly by its companion D and F plants, by the U, T and B chemical processing facilities, and by a fuel fabrication plant, sixty-four underground tanks for waste storage and assorted other complexes that supported Hanford operations. The plutonium was shipped to Los Alamos, first in cars driven by armed couriers, later by truck and by plane, and eventually in a specially equipped railroad car. In 1945 the construction camp was abandoned as the major building effort came to a close.

In many ways, the spirit existing during Hanford's war years would never again be equaled. Since blueprints were often completed just before each new machine system was to be built and were usually classified, supervisors relayed their specifications through oral instructions or crude drawings. Though this made each task more difficult, it also heightened the sense of being part of an urgent mission. Each job—whether milling the graphite blocks, bonding aluminum claddings on the uranium fuel rods, creating radiation sensing instruments or designing remote control equipment to extract plutonium from the spent fuel—offered challenges to ingenuity, craftsmanship and skill. Whatever the mysterious product being created here, its use would help defeat a barbaric enemy now threatening the world. (As an extra sacrifice, the Hanford men and women chipped in a percentage of their wages to finance an Air Force B-17 which was then named *Day's Pay*.) The work provided a sense of purpose far beyond that offered in normal daily life.

But while Hanford employees willingly worked their overtime shifts, put up with the slapdash accommodations and suffered through the two-hour daily rides on the buses which the project provided, they were frustrated not knowing what it was they were creating. Newly recruited workers were not even told where they were headed. The construction workers, as well as the engineers and scientists, worked on specific tasks and asked no unnecessary questions. Both groups were prohibited from discussing their jobs even with spouses or friends from other crews. Less than a fifth of the operations people knew the end product of their work, and most guessed it was bombs or munitions—or perhaps some mysterious superfuel derived from the thousands of graphite blocks brought in to build the reactor core. A standard joke was that Hanford was making "fourth-term Roosevelt campaign buttons." But when one worker took a graphite sliver into the mess hall and showed it to friends from another department, he was gone the next day—arrested by Military Intelligence.

M.I. also opened mail and listened in on long-distance phone calls to ensure no revealing information was released. They had Enrico Fermi come through under the name Henry Farmer and Arthur Compton under that of A. Comas to hide the nature of the atomic product. They checked the background of each Hanford worker and even classified the amount of beer consumed so spies couldn't determine the project's size by the number of employees present.

Monitors did warn the workers about radioactivity by timing them in "hot zones," measuring them for contamination and generally watching out for their safety. Rules prohibited eating in work areas, mandated the wearing of dosimeters to log radiation exposure and required medical inspection for even minor cuts that radioactive substances might have entered. Workers were forbidden to use the Boraxo soap which—because the boron it contained was the same as that in the reactor control rods—might contaminate the graphite and prevent a chain reaction from taking place. But the explanations of radioactivity were usually limited to vague comparisons with X rays. When an instrument team developed a radiation detection device they named Pluto, after Disney's constantly sniffing dog, project head General Leslie Groves decided it sounded too much like plutonium and made them change the name.

But if the security rules were frustrating, they removed from most Hanford workers the burden of judging the wisdom of what they were creating. Simply assuming that their efforts were necessary to win the war, they could immerse themselves in details. They could take pride in having overcome the harsh environment, the pressure and awkward restrictions to meet an unprecedented technical challenge. They had the satisfaction of carrying out a job they were asked to do.

On July 16, 1945, the world's first atomic explosion took place at Alamogordo, New Mexico, using a bomb made from Hanford plutonium. Since Germany had already surrendered and Japan never had developed a real atomic program, many top scientists (including Leo Szilard, future Atomic Energy Commission head Glenn Seaborg, Nobel Prize winner James Franck and Einstein) tried to have the weapon demonstrated in an uninhabited location rather than have it be employed against a human population. But the project, originally begun to neutralize a potential external threat, had now produced a destructive device so powerful that those who made the final decisions felt it would be folly not to use it to end the war. With a bomb made of Oak Ridge U-235 that fell on Hiroshima, and one of Hanford plutonium that was dropped on Nagasaki, the atomic era had its first public presence.

Perhaps because theirs was more exclusively a production facility than were the theoretical labs of Chicago or Los Alamos, most men and women at Hanford knew nothing of the debates over their product's use—and even those aware of what they were creating kept any apprehensions to themselves. Instead the workers heard the news of the bombs, realized their part in them and celebrated with cheering, laughing and champagne parties far into the night. No one considered until much later what other choices might have been possible. They knew only that their labors had brought to a close the most brutal war in history.

When the Manhattan Project ended, most Hanford people assumed the site would shut down entirely. But the Cold War began almost as the final bombs fell on Japan and the newly formed Atomic Energy Commission entrusted new operating contractor General Electric with building five more production reactors, two fuel reprocessing plants, and eighty-one waste storage tanks—all to generate plutonium for warheads now aimed at the Soviet Union. General Electric also built two thousand more houses in Richland to serve a population which expanded from 15,000 residents at the end of the war (the bulk of the workers left, of course, when construction was completed), to 22,000 in 1950. The atmosphere of security, mission and immersion in particular tasks became a routine way of life.

I visited Hanford in the summer of 1979, the fall of 1980 and again in the summer of 1981—to talk with the men and women who'd spent thirty-five years living and working first with atomic weapons and later with atomic power, and to talk with their children and with the nuclear migrants who'd come in recent years to build new reactors once again in the desert by the Columbia. While the external world had viewed the old hands initially as saviors and later as servants of warmongers and polluters, they viewed themselves as "normal" Americans. If they and the other nuclear proponents achieved their dreams, I wondered how much of their culture would become a common future for all of us.

Hanford today is already a model for the "nuclear parks" which, by concentrating all atomic operations in a few scattered sites, allow reactors to be built while distancing them geographically from environmentalist opposition. The Area, as it is called by most employees, now includes 570 fenced and guarded square miles through which 13,000 workers travel daily to their jobs in cars and in the old blue buses which carry them from Richland departure lots. The reservation begins in the north with B reactor and the seven other old production plants, abandoned since their decommissioning in the mid 1960s to early 1970s but still radioactive, which line the Columbia like a row of monuments from some future Valley of the Kings. Toward the westernmost of these (the plants were built at six-mile intervals for safety and security), the still operating N reactor produces both plutonium and enough electricity—

diffused over the Bonneville Power Association's five-state transmission grid—to supply the equivalent of half the yearly needs of Seattle. Six miles south, on a plateau halfway between Rattlesnake Mountain and the Columbia River, the "200 East" and "200 West" areas include the separations and processing facilities, plus acres of waste storage "tank farms" where the fruits of the initial reactors now rest adjacent to nuclear materials shipped in from Three Mile Island and other atomic plants across the country. On a basalt ridge between the two "200" areas test holes are being dug for an experimental project that may make this—already the largest waste site in the world (possibly excluding some unknown Soviet location)—a permanent repository of all of America's nuclear by-products. Farther southeast, fourteen miles across the sage and cheatgrass-covered sands over which roam elk, mule deer, coyotes and jackrabbits, an additional 9,000 construction workers are building three new reactors beneath pale skies piled high with brooding, intensely lit clouds. They are working for the Washington Public Power Supply System (WPPSS) utility consortium. The plants are now projected to cost $13 billion when completed. Including both WPPSS activity and that of the Hanford Department of Energy (DOE) contractors, no complex in the world employs more people, spends more annual dollars, or includes a greater diversity of nuclear activities than Hanford.* The reservation ends with the billion-dollar test breeder just to the west, and with the "300" area fuel production and research facilities a few miles farther south.

Inside Hanford's B reactor, the first built here, and the first in the world outside of the Stagg Field laboratory, a family-sized box of Tide sat on a concrete railroad loading dock. In front of the dock, a flatbed car carried a huge shielded cask. Rain leaked through the ceiling above, forming a thin screen of mist that filtered down past the red girders overhead. Except

*It is remotely possible some Soviet plant operates on an equivalent scale, but none of the American, British or French experts I have consulted are aware of any existing.

for this dripping the plant was silent.

The Atomic Energy Commission (AEC) closed down B plant in 1968: pulling its fuel elements, control rods and sensitive equipment; decontaminating its accessible surfaces through the same laborious scrubbing as was done to the "crapped up" room at Z plant; leaving concrete shield walls to guard a graphite core which, although all its fuel has been removed, will remain radioactive for far longer than human civilization has existed on this planet.

Now, nearly four decades after the reactor's beginnings, I stood on a wooden plank floor above 25 feet of water which once held spent fuel elements. When the reactor was operating, the elements—composed of Uranium-238 and a small amount of U-235—were loaded into the honeycomb lattices of the reactor's front. They were transformed and seasoned by the neutron reactions, like so many smokehouse hams, until their contents included the element plutonium, which, before its 1941 creation in one of Ernest Lawrence's newly developed cyclotrons, had existed on the planet only in minute quantities at a single West African site. A rear discharge chute slid the rods into the water. Workers picked them up with a claw-ended pole resembling the kind used by grocery owners to grab cans off high shelves, and placed them in buckets to be weighed on a wrought iron balance beam scale ("Honest weight," said a sign on its side) that now sat unused and rusting by the pool.

The workers used an array of overhead hooks, rails and gurneys to move the elements into numbered storage slots, where they rested for three months while the fiercest and shortest-lived radioactive isotopes decayed; then they loaded them into the black cask train that pulled up to the dock and took them eight miles across the desert sagebrush where their contents were dissolved, filtered and concentrated at the T reprocessing plant. Finally, with the separations completed, armed couriers brought the plutonium to Los Alamos, where it was shaped into the bomb that ushered in the nuclear age at Alamogordo, and the one labeled "Fat Man," which destroyed Nagasaki.

But the artifacts before me now could explain neither the

bombs nor what they brought to the world. What I perceived, rather, was a Shaker village, a haunted mansion, a cannery row warehouse containing an array of quaint objects all aged and weathered sufficiently to place them—were they just a little smaller—adjacent to the wrought iron scythes or wooden cart wheels decorating some hip young lawyer's urban living room. I watched Roy, a young United Nuclear Corporation worker, take the claw pole and try to open its water-submerged jaws. Ralph, the caretaker for all Hanford's decommissioned reactors, shone a flashlight through the water so we could watch the jaws as if they were those of a truculent crab. The tool stuck momentarily, then moved. After Roy used it to lift the handle of an old submerged bucket, he turned and announced with a grin, "This really blows me."

The reactor building was cold, and my United Nuclear PR guide spent most of the time shivering and wondering whether she wouldn't have been better off wearing socks instead of nylons. Ralph—who'd come to work during World War II to mill the uranium slugs—stood quietly in his battered London Fog. When he described technical specifics it was as if he were reminiscing about a friend to whom he wished to do the highest justice.

We entered the old control room next, passing one sign requiring a high-level "Q" clearance, another instructing workers to leave the area at the sound of an evacuation alarm and a third which depicted a rugged Superman with red and green swirls emanating from him and which explained, "Security is an individual responsibility . . . Be an individualist." The room's interior was simpler—far simpler than the control facilities for more modern Hanford projects such as the three WPPSS reactors, and even simpler than the nearby computer rooms where the waste storage tanks were monitored. B reactor had only an operator's console, some stands of roller-graphs, pressure and temperature monitors, and a few other measurement and control devices, plus a panel of hydraulic gauges with a notice cautioning that bumping the panel might cause a "scram," or automatic shutdown. We played here as well, pulling levers that would have withdrawn the boron control rods which absorbed mobile neutrons to slow or stop the fission process; rotating the metal expansion coils which—if they sensed pressure overload—would tip mercury vials, break circuits, and drop the vertical control rods; even lifting the hinged Plexiglas shield guarding a last-chance emergency shutdown system which would leave the reactor inoperative for months. "Isn't this amazing," Roy repeated, playing with each dial, switch and gauge as if he were manipulating the wing flap or turret gun of some childhood model airplane.

Returning to the loading dock, Ralph complained half-jokingly that he never got enough money to fix the 57 acres of roofs on the abandoned buildings for which he was responsible (concrete shield walls protected all contaminated areas), and talked with Roy about why the booming nuclear economy was making local rents unaffordable. Below us—where an overhead crane had dropped the lead casks down a remote-controlled wrench had removed their tops and another crane had loaded the highly radioactive buckets containing the elements—a light burned through the dark water.

With the end of the war, Hanford workers at last knew the product they had been creating. Turning their attention to peaceful applications, they discussed the possibilities of an atomic energy so limitless that users might not even have to meter it and debated whether it would be generated from the now wasted thermal energy of reactors such as B, D and F, or whether some yet to be invented process would enable electrons loosed in nuclear reactions to directly charge high-tension power lines. They received certificates signed by Secretary of War Henry Stimson thanking them for participating "in work essential to the production of the Atomic Bomb, thereby contributing to the successful conclusion of World War II," and letters from Du Pont's president expressing a similar message. With the newly formed AEC now the government body in charge, and with General Electric replacing Du Pont as prime contractor, Richland became known, to the local papers and many of its residents, as Atomic City.

At the same time Hanford's high-security atmosphere persisted. Engineers destroyed rough drafts, carbons and even typewriter ribbons used in preparing classified technical reports. They were still forbidden to talk about specific projects to their families or to workers lacking proper clearance. Billboards lining the road to the plants spelled out, in sequence, "Caution, Engage Brain Before Starting Mouth," "A Secret Can Circle The Globe Without Refueling" and "Alcohol Preserves Almost Anything Except A Secret." Later on Hanford's old hands would decide the restrictions had created a public eternally frightened about basically unexceptional technical processes. But with Klaus Fuchs giving away atomic secrets to the Kremlin, Winston Churchill warning about an Iron Curtain falling across Europe and columns in the Kennewick-based *Tri-City Herald* revealing how profession after profession had been exposed before the House Un-American Activities Committee for harboring Communists, Hanford's workers accepted readily the rules of silence.

To a degree, the very horror of Hiroshima and Nagasaki impelled unquestioning, unflagging efforts at Hanford. The A-bomb was the weapon which could have been used on us. It was the weapon which America's skill, vision and integrity had instead created first (or which the Lord had granted us, thought some of the more religious workers). We could allow no other nation to brandish a more powerful version of it against us.

Much of the work was only indirectly related to the ultimate "product." Find a more efficient means of circulating the cooling water from the Columbia. Create remote equipment for slicing open the fuel rods and safely extracting their radioactive contents. Develop new sensing instruments, temperature monitors, control rod mechanisms, fuel claddings and loading devices. Solving these varied challenges was always what J. Robert Oppenheimer called—referring to the actual bomb design—"a sweet problem," and that the scientists and engineers were pioneering new technological frontiers was as important as what the eventual product would be used for.

Although the normal work week dropped, after the war's conclusion, to a conventional forty hours, at times Hanford still demanded crash effort. One such case was when the Area was producing plutonium for the Eniwetok H-bomb tests. Although few workers knew they were contributing to anything beyond a conventional fission weapon, they started several months before, using blueprints drawn up by the Los Alamos theoreticians, to machine the plutonium hemispheres whose explosion would trigger the thermonuclear reaction. When the actual tests began, they labored around the clock honing the gray artificial metal to the finest tolerances possible, then loaded the "shapes," as the bomb components were called, onto waiting planes which flew them to the on-site assembly point. New results came in from each test. The designers altered the specifications for size, thickness and configuration just as they might have in any experimental process. Hanford's workers made the required changes, then flew the shapes out to begin the entire process once again.

Although the weapons heralded an age of unparalleled potential destruction, they were also mere mechanical devices. The reactors that produced their fuel were new machines to be pushed and developed to their limits. The Eniwetok preparation was, in the words of one of the men who actually did the machining, "just a job I did the fastest and to the tightest tolerances I knew how."

The men who founded Hanford considered themselves, in often-repeated words, "doers, not thinkers." That judgment had nothing to do with intellect—they were as savvy as any of their predecessors in America's long history of backyard inventors. But taking time to sort out the complex implications of their work would distract them from the building and creating they prized above everything else. They assumed their efforts fueled American progress toward increased strength and security, and they felt proud to provide for their families through good respectable jobs. They worried more about the practical questions of whether or not their machines would work than they did about how they would be used in the international confrontations whose ethics they left to the politicians and the preachers.

The old hands felt a joy in mastering the newly unleashed powers of atomic fission through an alchemical meld of parts, materials and purpose, a satisfaction in pioneering a desert once fit only for rattlesnakes and jackrabbits, and a sense of worth in creating working technical monuments that would endure long after the men who built them were gone. It was true that the nucelar stars—men such as Oppenheimer, Fermi, Szilard and Teller—were based not here but at Los Alamos, Oak Ridge and Chicago, and in part because of this Hanford never became as publicly known as did the other sites. But for all that theoretical foundations were developed elsewhere, it was in these reactors by the Columbia that nuclear technology became an industrial process, and that the men who manufactured plutonium, not in micrograms but in pounds and later hundreds of pounds, laid the ground for the massive atomic establishment America was soon to develop. Because the atomic industry would end up being staffed not by world-renowned physicists but by ordinary engineers and technicians, Hanford became the prototype for a nuclear future in terms of human as well as technical arrangements. To "tame" atomic processes was to bring them from the realm of the unexplored to that of the pedestrian and routine.

The old hands acknowledged this when they explained that their inventions "usually weren't like coming up with the wheel or the incandescent bulb." They were proud to "simply put a few things together and come up with ways to make existing machines run a little better." All changes seemed to be further steps in human evolution.

The desire to tinker applied not only in the labs and reactors of the Area, but in the old hands' basement and garage workshops as well. So when a retired engineer named Clark Reitnauer needed to heat-treat the guns he made as an amateur smith, he decided to build his own forge. He built it by connecting two automobile brake drums with a piece of pipe, attaching a bent metal rod for a handle and welding on a three-speed hand-held blow drier to provide the air supply. The forge now lay beneath a workbench in Clark's shop, a basement room with enough tools, machines and materials to fill a thousand and one nights of tinkerers' dreams. A maze of lathes, drills, saws and grinders covered the floor. Pegboards on every wall were weighted with pliers, hammers, wrenches, drill bits, calipers, micrometers and levels. Loops of copper wire and piping elbows hung from the ceiling like vines from jungle trees. Racks of baby-food jars held nuts, bolts, washers and screws beyond counting. The hot-air ducts that ran from the furnace were used as shelves to hold iron bars and stacks of orange and brown plastic circuit boards. Clark kept two beer cans sitting on one of these ducts; he had kept them for eighteen years, planning to incorporate them into a radio receiver, although he already had one two-way radio with worldwide range. He'd built it in 1959 from a $13 war surplus receiver and surplus circuit boards but hadn't used it much since then, because "I never got much kick from actually talking on it."

Clark Reitnauer transferred to Hanford in March of 1944 from Du Pont's heavy-water plant in Morgantown, West Virginia. A supervisor there had already taught him about high-level security by telling him, "One word about what we're doing here, and I'll have you incarcerated for the rest of the war." When his wife suggested, out of the blue, that Hanford might be making an atom bomb, he envisioned her telling people and himself jailed, then spent three days explaining how she was being ridiculous.

Clark began here building special planes and lathes to fabricate the B, D and F reactor graphite, then worked on a variety of radiation monitoring instruments, including the one abortively named "Pluto." Although he came to Hanford without a college background—he supplemented his high school education with night school and correspondence courses in mechanical engineering—Clark developed sixteen patents and was a senior engineer at United Nuclear by the time he retired in 1977.

As we stood talking in his basement workshop, Clark—in tan coveralls and corduroy slippers, his face stolid and rectangular, his neatly trimmed gray hair rising in twin peaks as if it wanted to soar off into a long, loose puff like Einstein's—leaned against an arc welder he'd built from a hand-wound transformer and scrap contacts and coils. Then he leafed with his weathered mechanic's hands through a sheaf of patents that bridged the gap between the megaproject world of the reactors and the isolated basements where individuals built better mousetraps while being affiliated to nothing except their own desires for invention. He explained how his ball-bearing rotator had been used to feed uranium slugs into an ultrasonic scanner that checked the bonds between the slugs and their aluminum claddings. How his induction heater realigned reactor tubes warped by neutron bombardment. How his "Japanese lanterns," with their six-foot-long steel lattice ribs, had been used to keep the thin aluminum tubes from breaking when they were pulled out by their ends. For Clark and the others like him, stringing together wires and pipes and switches was the very purpose of existence.

"Are you mechanical?" he asked, looking up from his patents. I told him I wasn't very, but he said, "I think you'll like this anyway." He showed me a picture of a young woman in a knit shift who was aiming some strange sort of gun while Clark stood, in overalls and glasses, giving benign counsel like some real-life Dr. Zharkoff showing Dale how to fight off Ming the Merciless. "We had to flare the ends of each reactor tube," he explained, "and since no one could think of a better method, we did it by hand, with cutting and bending tools. I said to myself, 'Hell's Bells, there's got to be a better way,' and, since I've always been an amateur gunsmith, I began by taking a surplus Springfield rifle and chopping the barrel down. I covered the end with a thick rubber nipple and placed a metal sheath over the nipple. When the gun was fired—without a bullet in it, of course—the exploding gases expanded the nipple and pushed out the sheath to flare the tube."

Clark described how he developed the device in collaboration with other engineers, then showed me other guns he'd rebuilt, like an old muzzle loader, worth $150 dollars but purchased "minus a few pieces" for $10 from a Kennewick junkyard. He described using his lathe to mill the missing hammer and the safety out of tool steel, heating and forging them on an anvil "just like they did in the old blacksmith days," then heat-treating them again on the brake-drum forge.

"I go a lot to the junk place," Clark continued. "To show you what kinds of bargains I get, I paid $5 for this slide I made into a Duwalt Angle Saw" [he showed me a special circular saw which rotated in all directions]. "I paid $13 for this $1500 Brown Monitor [a dark box mounted on the wall and containing a needle which traced on a roll of graph paper] that a junk guy thought was a broken intercom. I'll use it to control a new heat-treating furnace. I always say someday I'm going to take this mess and red it up—which is a Pennsylvania Dutch phrase that means to get things neat—but I never quite get to it."

When I asked if working with atomic reactions differed from basement tinkering, Clark said it was at first "a little mystifying and scary." But he adjusted quickly to entering hot zones so he could test the new systems he'd developed. He knew the dose limits were sufficiently low so no hazard existed. Even the year he was high man in terms of exposure didn't really worry him.

"I guess the monitors must have metered me wrong," Clark explained when I asked how this happened, but he said the annual dose allowable was normally three REMs and five as an absolute limit (REM stands for Roentgen Equivalent Man—a unit of radiation exposure that factors in the biological damage created by different radioactive emissions)—and that the dosimeter in his badge indicated he'd ended up with more than five. Since the international limit was fifteen, that didn't bother him. "But they had to file an AEC report, and they called me on the carpet just as if I'd broken the traffic laws by speeding."

Clark didn't consider his exposure any more dangerous than if he'd been driving 40 miles an hour in a zone marked 30. And even whether or not the zeal that created the violation served a wartime mission mattered less than one might expect.

"Of course when the boys stopped fighting, we cut our work week," he said. "We didn't have to worry any more that the Germans, who we knew had no morality, would get the mysterious weapon we were building and use it on us first. Now that all the goddamn bleeding hearts have taken over, what we did is supposed to be dirty pool. But we were proud to learn we'd made a bomb that saved us from losing thousands of lives invading Japan.

"When all that ended, it was almost no time before we were made afraid of the Russians." He thought maybe we always had to have a potential enemy; and he remembered how, along with so many others, he dreamed of nearly costless power. But mostly the men worked on their own particular projects and left the justifications to the top brass. For Clark it was still making machines work, and it didn't matter if they were optical systems for remote fueling, tooling for reactor maintenance or equipment for handling hot fuel sections. "I could just as easily have been working in a coal plant," he said. "They presented us with what they needed and we went out and built it."

The job men like Reitnauer did laid the technical base for an expansion program that ran from 1947 to 1953, and produced the reactors C, H, K East, K West and DR, the processing plants PUREX and REDOX and a support complex including shops that specialized in engines, instruments and boilers, glassware and cranes, fuel manufacture, optical systems and precision tools. They also built research and testing facilities and put together Hanford's own rail system with 150 miles of connecting track. The construction was performed by 16,000 workers who moved into a new camp north of Richland and lived in trailers and in naval barracks shipped up the Columbia by barge.

The Hanford engineers spent this period striving for greater production efficiency and gradually working their way up in the General Electric organization. Like Clark, many of these men lacked formal credentials: few who were wealthy enough to attend college in the Depression era worked at the power plants and shipyards that employed the old hands' previous talents, or chose this desolate location as a wartime project; and even those who were in school at the time often interrupted their educations for defense effort urgencies. But the Manhattan Project's crisis atmosphere permitted a class mobility that would have been difficult to achieve in more staid and settled times. Having been denied the longer-term, more abstract lessons of university education, the men learned on the job from their more experienced peers and from the trial and error testing of the machines they built. As their skill and responsibility increased, they received the equivalent of battlefield promotions which granted them higher rank, pay and even better housing built near the river in Richland's new northern section.

To a degree Hanford culture became more formal during the postwar years. While the top Du Pont managers made a point of visiting every work area and greeting each worker by name, GE brass concentrated on higher-level salaried employees. While Du Pont streamlined its operation with as few layers of hierarchy as were needed to carry out the mission, GE treated Hanford as just another division of a huge manufacturing corporation. Du Pont was highly paternalistic: Supervisors made it their responsibility to try to talk workers out of impending divorces and Clark Reitnauer once had to put his own position on the line to save the job of a kid who worked for him, "one hell of a good designer," who'd gotten drunk one night, broken into a grocery store and taken some beer. GE demanded equally proper behavior but was more tolerant in enforcing its mores.

The respect given to inventive prowess, however, was real, more than just an image in the ads GE ran in the *Herald* to announce abundant opportunity for "hard-working, honest, fast-thinking men and women." Project head Bill Johnson (in charge from 1952 until he left in 1966 to join the Atomic Energy Commission) was not only a dignified Englishman whose shirts were always immaculately pressed, whose suits were impeccably fitted and whose pearl tie tack was never out of place, he was also a tinkerer's tinkerer who himself held twelve patents and would wake up in the middle of the night to call one of the plants and insist that some system he realized was about to cause problems be checked and corrected. Since atomic technology was still experimental, few petty rules existed to hamper those who worked with it. If one took the plutonium product as a given, Hanford work provided a nearly perfect opportunity for building.

Socially as well as technically, Hanford people created their environment as they went along. Although stars like pianist Glenn Gould and singer Leontyne Price were periodically imported by Richland cultural organizations, workers demanding major league sports, ballet, theater or music generally returned to the urban centers from which they'd originally come. Those who stayed played on the local softball teams, joined the stamp, photography, gardening, skiing and kennel clubs that sprang up along with the Rotary, Lions, Kiwanis, Knights of Columbus and the Masons, and became both performers and audience for a Player's Guild, a light opera company, a group of meistersingers, and a forty-two-piece orchestra. They joined the town's twenty-five different churches (57 percent of residents attended, over twice the state average) where diverse regional accents made the choirs sound like Tower of Babel choruses. They attended art shows, dances and PTA meetings for the schools in which classes often held no two children from the same home state. They created over 250 community organizations for the less than 22,000 inhabitants that lived here in 1950. They used these activities to turn what was once an empty desert into a home.

Single Hanford workers lived in men's or women's dorms—actually almost apartments since they involved no parietal restrictions—and participated in the sailing, hiking and dancing events which the dorm clubs coordinated. Those who were married shared with their neighbors the excitement of raising their first kids—the town had a birth rate 20 percent above the national average—and pioneering a new domestic life. Because all these migrants left behind both families and previous friendships, they grabbed ready-made connections with whomever had shared the same home states, military units, university backgrounds or other projects previously worked on. They used these connections as a base until new friendships emerged from common Area projects or the new cultural institutions.

Despite the massive military enterprise supporting it, to its residents Richland was the atomic age equivalent of a homey small town. Since no one was allowed to live here except Hanford employees, their families and a few merchants running stores under government contract, Richland had no poor, no old and no unemployed. Crime was almost nonexistent—from 1945 to 1947 the local jail did not hold a single prisoner. The town newspaper—a nonprofit publication called the *Richland Villager*—ran from 1945 to 1950 featuring a regular column on the dorm club, announcements of whose relatives were visiting from out of town and who had been stopped for traffic violations and classifieds in which people offered to trade their "A," "H" or "R" houses for equivalent accommodations offering better access to the Columbia or to the town's newly built schools. Richland even had its own mascot, a jaunty potato-headed cartoon figure in overalls named Dupus Boomer (Dupus referring to Du Pont) who appeared each week in the *Villager*, chasing after the trash cans which the termination winds blew down the block, looking out at the desert while his kid asked, "Pop, how far away are we from the United States?" and joking with the local barber about all the "long-hair" scientists in town. That Hanford's workers considered themselves "a fine class of people" testified not to any snobbery, but to optimism and innocence.

The designers building the original Hanford reactors were uncertain whether the plants would indeed produce the desired plutonium and if they did, whether the planned atomic weapons would work. By the 1950s, with this production assumed and with an ongoing demand from America's rapidly expanding nuclear weapons arsenal, their goals were to extend efficiency and working life in what had shifted from an experimental to an industrial process.

One typical problem began with the reactor core: After a half dozen years of operation, the graphite blocks started expanding and began to warp the uranium tubes that ran through them. Although higher operating temperatures would prevent this from happening, the existing control rods melted too easily. They called John Rector, who was known as a troubleshooter, to design the new rods. Rector came to Hanford in March of 1944 as a machinist, after making dies for 50-caliber machine gun ammunition at Du Pont's Kansas City Remington Arms plant. He joined Clark Reitnauer in milling the original graphite blocks and worked at the T and U plutonium separation plants designing equipment from the rough sketches his superiors gave him. He advanced to engineer's rank through later inventions such as a special attachment called a vacuum chuck which used suction to hold the plutonium H-bomb hemispheres and mill them faster and far more precisely. (Although Rector's superiors didn't think the chuck would work and wouldn't assign him to build it, Rector convinced a sympathetic foreman to write him up for two weeks as if he were working on another project, while John used surplus lead to create the successful prototype.) He believed the reactors were necessary so America could stockpile plutonium "just the way we'd stockpile gunpowder if we were defending ourselves with rifles."

Rector began the control rod project by checking metallurgical studies at the library. The existing rods used an aluminum core with boron sprayed onto it to absorb the neutrons and halt the reaction. Could a more heat-resistant metal be combined with the boron? The studies said neither nickel nor stainless steel nor aluminum would mix in the right proportions through conventional approaches of melting together and alloying. Rector had checked nearly every possibility when he read of an obscure process involving compressed and heated powdered metal.

"It was called sintering, and had been used mostly to make automotive bearings," Rector explained, smiling in recollection as we sat in his office at the company, Western Sintering, which he ended up founding ten years later using machines he had hand-built in his basement. He was constantly sketching and diagramming with the three pens that lined the pocket of his open-necked blue shirt. His office—with its wall charts of metric equivalents, drill sizes and metal specifications, its file cabinets filled with orders and price quotations, its shelves lined with every technical manual a metallurgist could need—was a working mechanic's nearly ideal ready reference room. Rector began rummaging through his desk drawer—cluttered with Band-Aids, masking tape, colored pencils and odd nuts, bolts and screws—with the annoyed air of someone who's lost something.

"Wait, here it is," he said, and handed me a two-inch-wide ring of what looked like gray plastic.

"That's metal?" I asked. And he explained how he mixed powdered boron carbide with powdered aluminum, shaped the mixture using a press and a mold, then ran each ring through a furnace that bonded together the particles.

"I'd never heard of sintering before," he said. "It had only been used with iron, bronze and brass, not special metals like boron. But I kept playing until the mixture hung together and we could string 120 rings like so many beads on an aluminum test rod. The rod worked perfectly and took all the heat we could give it. We contracted production to a Los Angeles company that had been sintering since the 1930s for Chrysler."

A bit later we headed out for lunch in the new brown Mercury he called "my truck." The back seat was filled with an array of wrenches, levels, hammers and boxes of nails reminiscent of Clark Reitnauer's basement, plus a pair of jeans and a sweatshirt Rector wore for his own weekend tinkering. He explained, as we drove to Richland's Hanford House motel, how most architects "just pad their percentages by making buildings unnecessarily fancy. Take schools," he continued, using an example of waste I'd hear repeated almost word for word on two separate occasions here. "You could work out standard designs for the best classroom size, the best lighting and everything else. If local people wanted to add on at their own expense they should be able to; but instead they spend our public money funding different designs so every architect can have his personal monument. The same's true for airports. A maintenance man in Kansas City told me their hardwood floor cost thousands of dollars a year because they swell every time the roof leaks and they get wet. If you've ever seen those twelve-foot-high doors at Chicago's O'Hare, you just know they're there because somebody thought they'd be impressive. If people would only realize how easy it is to be cheap and functional maybe we wouldn't have to work the first six months of every year just paying off taxes."

Hanford House, Richland's star motel, was located on the town's main boulevard, George Washington Way, across a perpetually unused grass mall from the Federal Building's white bureaucratic battlements. Its two-story exterior resembled a circular slide carousel done up with alternating purple and turquoise concrete panels. Its restaurant sported square purple lanterns like some futuristic stage props for *The Mikado*, a fine picture-window view of the slowly moving Columbia and a large wooden planter with dry brush at each end and chrysanthemums down the middle.

Resuming his narrative with the knowing grin of a machine prophet long tested by technological infidels, Rector recalled how the Los Angeles sintering firm delayed manufacturing the rods until the deadline date was extended first by four, then eight, and finally by ten months after the order was originally due. "So my boss sent me down as a consultant, but since I didn't have command power, I waited three months more while they tried every wrong method in the book, then ended up producing the rods exactly as we had back in the lab. For all their mistakes, they still made money, and that made me think sintering might be a good growing field to get into."

Rector's desire to build fast and efficiently was akin to that of nearly all Hanford old hands who wanted to get the job done without the interference of incompetent outsiders like the Los Angeles contractor, the harassing government inspectors or the higher-ups who'd nearly stopped John from producing his vacuum chuck. Wartime urgency had provided the perfect climate for unquestioned, full-speed-ahead progress. Although the spirit of common effort and pioneering enterprise had not been equaled since, it remained an ideal cherished by all here.

If Rector was a tinkerer in a culture of tinkerers, he was also, in contrast with many here, an individualist who wanted to produce his own product and be his own boss. So while most Hanford workers—unready to "leave something safe and risk everything they'd made and everything they owned"—remained company men, Rector researched what he would need for his own sintering operation.

"Unfortunately," he said, "even for the cheapest used models, the required furnace would have cost ten thousand dollars, the press seven thousand, and the ammonia cracker for the furnace three or four. Those were millionaire figures back then. And since the banks, as usual, only lent to those who didn't need it in the first place, I had to keep my Hanford job."

So Rector continued at the Area. He put in his eight-hour

days, fabricating weapons-grade plutonium which, since the secret was over but the security state was not, was shipped to Los Alamos in a converted hospital car (called the "Round Robin") and sandwiched into regular trains behind the engine, equipped with its own cook, its own mechanic and armed guards who rode along for the journey. He designed new production systems and felt proud that his efforts were protecting America. But while Rector gave his Hanford job "my highest priority," he also began making payments on an $1800 lathe and a $180 welding kit. He read everything written about sintering. He scrounged junkyards for parts and scrap steel. For five years he put in the equivalent of another full shift plus every spare dollar he had, and, milling parts where necessary out of raw iron bars, he built the required machines in his basement.

"When they were completed, that was, of course, only half the battle," Rector said, finishing his French Dip sandwich. "The businesses I solicited still all asked, 'Well, who have you sold to before?' Even if Hanford projects would have bought from me, they needed only the control rods they already had. It took five years from when I made a $78 sale to a Kennewick potato bagger until I finally built up enough business to quit my job. Then I raised $100,000 from friends and $120,000 from a Small Business Administration loan which almost wasn't worth the year of harassment they put me through. I bought the land, built my building and began full manufacture."

When we got back to the plant, an overalled employee approached John and, describing an alignment difficulty on one of the lathes, explained, "We can't slot the plate in."

"Oh yes, you can," answered Rector, drawing a quick descriptive diagram and beaming like an eight-year-old because he knew exactly what to do. "You put a channel on top, these eye bolts over there, and you can slot your unit right above." The man went off to build the mechanism. Rector took me inside the shop to see the half million dollars' worth of equipment he now owned.

As the radio played the disco song "Le Freak, C'est Chic,"

and as the stamping machines overlayed a syncopated bass and drum accompaniment, John showed me the original press—still working along with his hand-built furnace and hand-built ammonia cracker. Powder descended from a hopper through a corrugated rubber tube and a hollow movable bar pushed it, as if it were dirt swept by a broom, into a recessed die. The press was producing copy machine rollers, and as I picked one up and circled it neatly between my thumb and forefinger, Rector explained the machine's capacity was fifty tons, "because that was the size of the surplus tubing I found. It could have been forty, I had wanted it sixty. The size actually turned out lucky because you use almost the same mechanisms to compress uranium oxide into reactor fuel and since the press is small enough to fit inside glove boxes, I've sold over forty slightly reworked models to nuclear sites around the country." He showed me one in the adjacent machine shop, waiting to be shipped with its $40,000 price tag to Los Alamos. He guided me past the 2,000-degree furnace, the conveyor belt that ran through the furnace carrying rows of green bar washers for the leaf springs of semi-trucks, and past a fourteen-foot-high rack which loaded the belt automatically, using an electric eye monitor, allowing it to run all night while the employees slept soundly at home. The loader was Rector's invention; he built it himself—along with a Rube Goldberg contraption of cams, gears and trip levers which removed the heat-treated parts from the furnace's far end—"because I wanted to buy the equipment and no one made it."

A friend and former co-worker of Rector's once called him "an Ayn Rand hero." When I asked John earlier if the wide open spaces of this desert country aided dreams of building and pioneering, he smiled and answered, "I think it's more what you have inside you." Whatever "it" was (perhaps the tinkerer's equivalent of "The Right Stuff" Tom Wolfe found in the test pilot astronauts), Rector had it. He got his education Horatio Alger-style through reading, experimenting and attending technical conventions where he could learn from the pioneers in his field (since the American Society for Metals elected him president for the 1956–57 term, they obviously thought he could return the favor). He threw his entire life into the work that first took two shifts and later a single one of one hundred hours a week. He built his company literally by hand, and remained friendly enough to stop and chat along the way with fellow inventors. Like Reitnauer, like everyone who'd come here in a context furnishing unlimited urgency, unlimited funds and an unlimited faith in invention, John seemed at times more than human in his ability to create and tend machines. I thought of Mark Twain's Connecticut Yankee, bringing the telephone, indoor plumbing and the Gatling gun to the medieval England of King Arthur's court, and of the sorcerer's aura that accompanied him; watching these men invent tools and tool systems made me almost accept their belief that all problems had mechanical solutions.

Rector roamed the plant, adjusting a screw here, a pressure monitor there, jumping in to help tug a drill housing back in line, and wiping his hands with a back-pocket rag after each encounter. Machine-building was simply the way he approached the world, and I envisioned him building his own airplane, as had others here, inventing the new ultraefficient windmills for which he'd come up with a secret design, even—had he been so inclined—fabricating a homemade atom bomb in his basement. But Rector was a peaceful man for whom bombs—and the reactors that generated both the plutonium to arm them and commercial power—were simply machines to play with and make more efficient. He assumed the government and corporations which contracted for them had their reasons for requesting production. It was fine that others planned and he built.

Rector checked the belt feeder once more before leaving the plant, explaining, "It don't move too fast but it don't have to." He said that, assuming the weather was nice, he'd spend the coming weekend working on a pumping system to supply water to an auto racetrack he had built on the outskirts of town:

"So if you'd like to visit while I putter and relax, laying a bit of wire, a bit of pipe and a bit of conduit, you're more than welcome to come on down."

The fact that Rector had his own manufacturing company made him an exception in what was and still is a company town. Richland had its clothing, drug and sporting goods stores, its supermarkets, its gas stations, and even two jewelers and a Studebaker dealer—all divided between the community's original business district and the new Uptown Shopping Center near the north end of town. But aside from the Area, there were no industrial enterprises larger than a dry cleaning plant. Businessmen had to go through complex procedures to get permits and to pay out a portion of their proceeds to GE. As of 1950, the town had less than a fourth the number of retail establishments of similar-sized Washington cities.

But residents needed few local entrepreneurs when GE supplied free lawn seed and topsoil, plus sapling trees to line their sidewalks, and when Tenant Services came, during the war years, to change light bulbs and fuses, fix clogged toilets or replace torn screens. The rents remained a bargain: $37.50 a month in 1949 for a three-bedroom unit in one of the two-story duplexes designated A houses, $33.50 for two bedrooms in a single-story B duplex and $50.00 for the single-family three-bedroom H model ranch houses, which served as early prime accommodations until the Q's, R's and S's were built. The major problem was when new residents came to town and had to live in the men's or women's dorms until their names moved slowly up the 600-person waiting list that was posted by Tenant Services on an outdoor bulletin board.

Although the houses were only theirs as long as they worked at the Area—those who lost their jobs for any reason had to leave—the Richlanders soon began fixing them up. They planted sycamores, firs and beeches; the trees grew to shade the houses and were regarded almost as symbols of a community created through its own efforts from a barren desert. The inhabitants surrounded the trees with perfectly manicured lawns which they had to rake and reseed each time the termination winds blew dust back over them. They enclosed their lawns with white picket fences and complemented them with pristine rock gardens bordered by red and white geraniums. As a *Villager* editorial explained, the well-groomed yards were "truly reflective of the character of people who live here."

Making the houses livable was harder. A sometimes ignored law on the books from 1946 to 1950 prohibited the building of garages or sheds and any type of floor refinishing except waxing; detailed instructions even explained what one could or couldn't do in terms of installing air conditioners or adjusting the basement furnaces. But the master tinkerers did what they could despite all obstacles. John Rector and Clark Reitnauer cleared and floored their half-excavated basements so they'd have space for workshops. Another old hand put in a grape arbor, which his neighbors said was foolish until he ended up eating the fruit just a few years later. Others did whatever else they could to create environments as comfortable and normal as those they would have found anywhere else.

Of course, some Hanford workers—around 20 percent as of 1950—found it uncomfortable to live in a town run in every detail by their employer. They wanted—even if it meant they had to pay utilities, school taxes and mortgages—to raise animals, to live adjacent to neighbors of varying occupations, "to own our own houses in a real community." They moved largely to the nearby towns of Kennewick and Pasco, along with those who sold goods and services to the Area and its workers, and along with the retailers and small manufacturers who served the farmers growing alfalfa, wheat, potatoes and sugar beets in the surrounding Columbia Basin Irrigation Project. Kennewick—seven miles southeast on the same side of the Columbia, past the Yakima's marshy flood plain—grew from a 2,000-person 1940 farm town whose three-block downtown strip included a bank, a hardware store, several drugstores, groceries and cafés, the local barber shop and some farm equipment suppliers to a community of 10,000 in 1950 and 14,000 ten years later. Pasco—an old railroad center which developed across the Columbia, a few miles farther south than Kennewick—grew from a town of 4,000 when Hanford was built to one of equivalent size. Although Pasco's proportional growth was less, it began with a greater agricultural and industrial base, and remained more an independent entity than a bedroom community for Hanford.

For most Area workers, however, Richland was home or, more properly, became a home as new ties developed and old ones faded. Even those who saw the town originally as a sort of Grand Hotel, which the government had created and could take away at any time, settled in—going to work every day in a familiar place, greeting friends with whom they now had a history and coming home to streets increasingly overlaid with memories. Their kids went to school, grew up and often later got Hanford jobs and raised their own families here. The trees and lawns grew, as did a shelter belt of Russian olives, black locusts, sycamores and various evergreens that was planted west of town to shield against the sand, the grit and the termination winds. Relatives and colleagues back home died, dropped out of contact or moved on to new communities themselves. Both the technical and cultural projects began to exert an ongoing pull.

The settling in wasn't universally embraced. For all those who, like John Rector, saw no need to think about places left behind, many others remained wistful at each of the Iowa, Indiana and New Jersey Day picnics that took place (and, for former residents of most states, still take place) every summer. In a 1952 League of Women Voters survey only half the respondents considered Richland their permanent home. But just as Hanford's plutonium manufacture became routine, so Richland slowly shed its makeshift character and began striving, almost like Pinocchio, to become a real town in the mainstream world.

At times this desire took the form of strutting. Atomic Frontier Days began in 1948 as an annual Western-style celebration. Movie stars visited, the men put on fake beards and held a male beauty contest, and local organizations used wire, crepe paper and paint to turn cars and trucks into elaborate floats. A diaper service built a huge winged stork. Mother Hubbard and her children proclaimed the merits of a shoe store. Members of Rainbow Girls dressed in an array of spectacular hues. Judges picked Miss Richland, one year selecting local belle and future Hollywood actress Sharon Tate. On a more serious patriotic note, the Navy's Blue Angel jets performed acrobatics overhead. A ground parade showcased tanks, howitzers and Nike missiles from the protective base on top of Rattlesnake Mountain; the high school sports teams, the Richland Bombers, rode by in their yellow and green colors displaying a finned metal bomb.

Most of the changes, though, required more responsibility on the part of Richland residents. The AEC and GE supported this because total paternalism, despite the social control it facilitated, was a bureaucratic headache. Town civic leaders wanted more independence because they dreamed of general economic expansion and because there seemed something almost un-American about what was now a civilian project still being run like a military base. The reactions of most Richlanders were mixed.

The shift actually began in 1948 when the government sold to town residents the furniture it had supplied free during the war years so they wouldn't have to waste time shopping or waiting for their own to arrive cross-country. The next year tenants had to replace their own fuses, faucet washers and broken windows. In 1951, as the Chamber of Commerce pushed for Richland to become a private city, GE announced a major rent increase that would bring the town's housing costs in line with those of equivalent accommodations elsewhere. But as all Richland residents worked for Hanford, the local unions considered this a *de facto* pay cut, organized a massive letter-writing protest to Congress and managed to keep rents at their previous level.

Gradually Richland moved toward becoming a normal single industry town. Electric meters were installed, then water meters. A 1955 advisory ballot on self-government lost by 500 votes. The AEC decided to sell the property anyway, for 50 percent of appraised value, and in 1956, 1500 residents gathered at the Bomber Bowl to protest the appraisals running too high. A delegation flew to Washington D.C. to work out compromise prices. The houses were finally offered at bargain rates: the mortgage on a $9,000 Q model, one of the best available, ran only $90 a month; H models went for $6,000; one could buy both units of the A duplexes for $7,300 (they were offered first to the family that had been there the longest; the other tenants could still remain as renters). Purchasers could even arrange for a guaranteed buy-back if the Area folded, although most here took the 10 percent discount that went along with waiving this option.

Not everyone jumped at committing themselves to the community: Clark Reitnauer, for instance, came close to leaving and going to Alaska, and a number of workers preferred having at least some alternative economic base and consequently chose to buy or rent in Kennewick instead. But the mortgage payments weren't substantially higher than previous rents. Owning their own homes allowed the Richlanders to fulfill the American ideal of each family possessing an independent freehold, and it allowed them to keep their houses if they retired or took other jobs. For those who dreamed of developing an atomic metropolis, establishing some private turf was a prerequisite.

To accompany this shift, GE and the AEC withdrew as well from administering municipal services. A bill passed the Washington State legislature allowing Richland to become a major Class III city without passing first through the conventional stages of Class I and Class II. The AEC set aside funds for street improvements, a city hall, a second fire station, library and hospital and a variety of other institutions to fulfill functions previously provided by the company. In December of 1958, Richland became its own incorporated town at a ceremony attended by Governor Albert Rossellini and Senators Henry "Scoop" Jackson and Warren Magnuson and capped by the setting off of a mock atomic bomb.

2
Mom and the Kids

"A woman is a production (we say a *re*production) unit. Whenever nature is faced with the construction of a new individual she quickly converts her parent plant from peacetime to a war activity, as it were."—from *Teen Days*, a 1946 guide for adolescents

While Hanford's men were manufacturing plutonium, their wives had the task of bringing civilization to Richland's former desert streets. They arrived in a period of wartime shortages: ordering clothes from Sears and Montgomery Ward because Richland had only a single, poorly stocked department store, and pooling their weekly gas rations so they could visit surrounding farms to buy eggs, vegetables and fruit. Their houses were so identical in looks that they would occasionally mistake those of their neighbors for their own. Their husbands, men like Clark Reitnauer and John Rector, were formally banned from discussing the work they were doing.

They were cut off geographically by the government-planted shelter belt to the west, Hanford to the north, the Columbia to the east, and the Yakima River marshes blocking all but a thin strip of land leading to Kennewick and Pasco. With Spokane 100 miles away and Seattle or Portland 200, with the Tri-Cities on the path of no major interstates and the Richland and Pasco airports barely allowing twin-engined planes, the isolation produced by their husbands' silence was compounded by that of living—as if in a colonial compound—surrounded only by other newly arrived Hanford wives.

Despite these difficulties, the women here—like those who

spent the postwar era pioneering suburbs across America with names like View Ridge, Pleasant Hills and Sunset Gardens—felt they could bring white picket-fenced domesticity to this randomly formed community. They compensated for their husbands' immersion in concrete, tangible machines by fitting themselves, even more than most American women of the time, into the roles of decorators, comforters and nurturers. They packed bag lunches for their men every morning (the on-site mess halls closed once plant operations began), and they dusted and cleaned to make their homes immaculate relaxation havens. They baked cookies for the neighborhood kids and answered politely when the gray-suited FBI men came around each year to ask whether their neighbors drank too much, spent too much money, partied too late, carried on illicit liaisons or did anything else that might leave them vulnerable to the blackmail of spies.

Because professional entertainment was scarce and attending store openings or riding around town on the free Richland bus system hardly satisfied cultural needs, the women, like the men, ended up creating their own local institutions. They spearheaded the performance groups, formed garden, bridge and kennel clubs, and worked with the PTA, the Brownies, and the Kadlec Hospital auxiliaries. They joined churches like the Central United Protestant congregation (whose motto was "Where the atom is split the churches unite"), sponsored youth groups and raised money for overseas missions. The local papers wrote up organizational meetings. Partly because of this and partly because they had no other outlets for style, the women at these gatherings poured tea and coffee from silver urns, wore their finest clothing and used china plates to serve the cherry cheesecakes and lemon meringue pies they'd baked.

A present-day bridge club I attended met every two weeks, with the job of hostess rotating among its eight members. It was held this evening at the house of Virginia Dumont, a Hanford wife since the early days who had become involved, at age sixty-five, in the women's movement, and who now divided her time between her longtime peers and young feminist friends. I asked the women about the wife of Army engineer Colonel Henry Kadlec, who had been in charge of building the town. The Colonel himself, I'd heard, had been essentially good-hearted, but "Lady Kadlec," as she was generally called, was notorious for cutting in front of the two-hour lines at the bank, grocery, gas station or movie theater whenever she felt like it, for requisitioning laborers off plant construction to fix up the custom-remodeled farmhouse she moved into and the riding academy she deemed an essential part of civilized life, and for attempting to fire all who talked back or got in her way.

When I asked if these stories were true, a curly-haired, enthusiastic PTA mom named Ethel shook her head, placed her hands on her flowered blouse, and said, "Oh let's not talk about the bad things . . ." The group debated momentarily whether the higher-ups ever did pull rank; Virginia recalled how, even after the Colonel died in 1944, his wife always demanded a special table at the officers' club, and the other women responded that her request was only just, given her position. Ethel changed the subject to Richland's excellent athletes, one of whom had made the Pittsburgh Steelers and one the Dallas Cowboys, and to all the presidents who'd visited—Eisenhower at McNary Dam, JFK when the N reactor opened, Johnson at Ice Harbor Dam and Nixon at Battelle Corporation's research labs.

"The dams aren't part of Hanford," said Frances, a transplanted Tennessee belle wearing a brown polka-dot jacket.

"Well, they're close enough," responded Ethel, who went on to describe how finding Indian arrowheads in the hills made her feel "part of a plan."

Frances told me about Atomic Frontier Days, and then recalled: "In any other place this size, if my daughter was going to the movies or even a store she'd have to pass a slum or rundown area. But here we have nothing like that."

"When I didn't let my twelve-year-old daughter walk home alone from a movie, she told me, 'Mom, you've been watching too much TV,' " said Doreen. "But actually, this was both a safe place and one where our kids could grow up without the slightest prejudice against any race, color, creed or position."

"I didn't realize there were any blacks in town," I said.

"Yes there were, the Brown boys. We treated them just like everyone else, and did you know they did a survey and found there was absolutely no prejudice at all here?"

Letting the subject drop, I asked instead whether their husbands talked with them about work once the security eased.

"I knew absolutely nothing about what mine did," said Ethel.

Sue Pearson, a sweet shy grandmother who liked to recall how her friend saw a woman in a brand-new Cadillac buy groceries with food stamps, nodded her head in agreement. A woman named Sarah explained: "I wouldn't have understood if mine had. He did talk about personnel problems because he was in waste disposal and worried a lot about safety and on-the-job carelessness. He wanted everything to be . . ."

Sarah stopped, fumbling for the right word. Barbara Giroux, who'd been quiet so far, interjected "according to Hoyle." The group laughed at the bridge reference.

The remaining women agreed that atomic work was never discussed. Doreen said she "didn't want to know." Virginia recalled how the kids in her 1960 third-grade classroom knew only that their fathers worked "in an office," then continued: "In my own case, my husband Lester would come home from work announcing, 'I'm really . . .' "

Virginia stopped, about to say "I'm really pissed," but checked herself because this was a group where, even when quoting someone else, one neither cursed nor criticized the Area world with harsh words. "Well, he didn't say it exactly that way," she said with a smile. She made it clear that her husband discussed both his discontents and his successes with her.

This made Virginia the exception, as she was in many ways already. But among the other women, it seemed only Sarah had any hint of what her husband did each day. Because I was still unsure how much she knew, I asked her: "Exactly what safety difficulties did he tell you about?"

Before Sarah could answer, Doreen jumped in angrily: "You asked about safety? I'm worried you're like all the newspaper people. That's all they're ever interested in talking about. There never have been accidents here. It's completely safe. But all people think about is Hiroshima, and they don't realize that even when the bomb was dropped it was already curing cancer. We had a class in Sunday school discussing how God gave men the atom for peace. I feel we've been given a directive from somewhere on high to continue instead of turning things over to quacks from Seattle or New York."

"That's right," added Ethel. "Energy, we need energy," and Doreen continued by attacking Jane Fonda and people "who don't know anything but try to tear us down."

Ethel reminded me of the Central United Protestant "Where the atom is split . . ." motto. She recalled when her husband spoke at a California church "who'd already sent people to march against the atom," then stopped for a moment because the conversation was getting heated. "Isn't it fun to reminisce," she said with a disarming smile. "But I just get on a soap box, because if we'd told our kids more about how important the work was, they'd have been prepared when all those professors started criticizing it."

If the codes of Hanford's domestic world were less formally articulated than those of the actual atomic reservation, they were nonetheless adhered to strictly. Broach the issue of safety or appropriateness and you threaten the entire project. Threaten the project and you jeopardize not only concrete buildings, stainless steel pipes and zirconium-clad fuel rods, but also the Richland of flower gardens, cleanly waxed floors—and thirty-five years' worth of friendships.

Hanford women accepted their social constraints in the same casual manner they accepted the implications of what their husbands produced. If the bombs existed, there had to be a reason. If the Area was the sole support of the town, was that any different from having the paychecks signed by Proctor and Gamble in Cincinnati, by General Motors in Midland, Michigan, or even by Du Pont, most anywhere in Delaware? True, a mother worried if her son was in a fight and the other boy's father was a top-level manager. If you wanted to leave the home to work, normally supportive networks would, even more than elsewhere, begin to condemn you. But if the contracts kept coming and the husbands were securely employed, why worry?

"It was important to be part of the right circles," said Virginia when I arrived at her house to talk further. "When we were invited to afternoon teas or sherry parties we'd check in advance to see if the other women were wearing gloves or hats. We felt whether we were tastefully and attractively dressed would reflect on our husbands. I remember I was playing bridge one time with Marion Prout—her husband was the top GE manager before Bill Johnson—and she doubled me when I overbid. I went down eight hundred points and came home thinking I'd blown Lester's career for sure."

The responsibilities faced by Virginia and the other women paralleled to a certain degree those of their constantly inventing husbands. As they labored to make their homes spotless, their children well-bred and themselves attractive and cultured, they transformed Richland from a crash military project to a normal domestic environment. With Hanford the town's sole employer, their status within the hierarchy of this environment was always clear, and they felt it an achievement to associate with women whose men ranked just above their own.

When Marion Prout switched her women's auxiliary efforts from the Orthopedic Guild to Kadlec Hospital, many of the wives followed suit. When Barbara Giroux's friend Joan divorced and remarried a man of lower Hanford standing, their next-door neighbor said they no longer had to see her any more, and Barbara's own husband rationalized cutting off relations by saying "he's just not very interesting."

"In a way the parties were part of a job," Virginia said, "going places to be seen and make contacts. But when we were stuck in the house dusting, preparing chicken and washing our husbands' socks, they were also a welcome outlet. I tend to laugh now at the meetings where they announce each time whose tea service we're using, but back then it seemed an honor. I think the feminist projects I'm involved in now are probably more important, but it was still useful to run the hospital bookmobile, raise money for costly equipment and change the flowers in patients' rooms. You have to remember we grew up accepting that things were to be a certain way and had no realization they could ever be different."

Virginia was raised in Utah and Idaho mining towns while her father, a failed dryland wheat farmer, moved from place to place, taking whatever jobs he could find as a butcher, a mine guard, a painter and a handyman. She learned to read early and always carried a few favorite books with her as she traveled. Once, when she accidentally dropped a copy of *Black Beauty* from a train window, she cried for a week.

In Chicago, Virginia decided to be a teacher and eventually worked her way through Northwestern. She was poor most of the time and had to borrow $52 one semester from a man she was dating, which made her feel almost like a prostitute.

When she came to Hanford with her husband in 1950, Virginia slipped immediately into the role of housewife and mother. "I had the ring and the eligible man," she said, "but not working left me depressed and restless. I mentioned this to my husband and to my own mother, but they assured me it was something that would pass. I tried teaching again in 1955, but my daughter fell out of a swing and suffered some minor scrapes. Even though it was nothing serious, I was so filled with guilt about 'Oh, if I'd only been there,' I ended up quitting again for another five years."

Virginia's eventual return to work meant she withdrew somewhat from the auxiliary, bridge club and party circuit, no longer had time "to make friends with the wife of so and so who was a hot new administrator or scientist," and missed some of the caring community she valued. But it brought her in contact with a new circle of women "who actually earned a separate living." She joined curriculum committees and took workshops in how to teach reading, math and geography. She came in contact with imperatives other than those of the Area and its culture.

It was partly from teaching, Virginia explained, and later from a consciousness-raising group she joined for a year in 1970, that she began to sense that "all the nurturing and fostering roles" she'd learned her whole life were in some ways limiting. Why did her friend Dorothy have no idea how much money her husband had? Why did all her other friends "end up saying 'Well, I didn't think buying that chair was a good idea, but it's his choice, I guess,' as if all financial decisions were automatically the right of the men they were married to?" She said, however, that she too was a part of this place, and it seemed clear she'd learned the rules for setting a proper table, being a skillful hostess and dressing right—even if the deep reds and oranges she wore evoked a different feeling from that of the demure whites, pinks and grays preferred by her bridge club peers. She remembered thinking the FBI visits were necessary to guard the atomic secrets, and she raised her kids and joined the Orthopedic Guild just like everyone else. "But I'd come back all excited from a consciousness-raising session, and Sue Pearson—a sweet and well-meaning woman if there ever was one—would always change the subject by saying 'Well, I've heard that before, and you know it's only the disappointed ones who become women's libbers.' "

In some ways Virginia's husband Lester was also a maverick: fighting constantly for higher quality standards on the fuel element claddings that it was his job to check ("Sometimes he was right and they didn't work, sometimes not and ones he'd considered shoddy would be used with no problem for years"); being passed over for promotions and shunted into the role of consultant instead of line management because he spoke his mind and was abrasive; being pleased when Virginia returned to teaching and taking her seriously enough to explain his work instead of guarding it as a "man's world" mission. Still, he was a Hanford engineer, and when his overzealousness froze his salary and status he'd apologize for not rising higher. When Virginia raised questions, in recent years, on waste disposal safety, he'd dismiss them by insisting simply that there was "nothing to worry about."

In part because Lester did respect Virginia's independence far more than most Hanford men would have, life for her was by no means entirely constrained. She had her teaching and his endorsement of it. She had the neighbors with whom they shared lawn mowers and tools, and swam and drank together on a dock the men built. (She still swam during the summer, and a young woman she knew remembers her coming out of the Columbia with the water dripping off her short brown hair and tanned arms, while several men, old friends, stared without realizing they were doing so, and forgot entirely that Virginia was seventy-five.) She and Lester even bought a boat together with Sam and Henrietta Beerman from next door—and this last led to a friendly rivalry during the 1960 election, after Democrats Sam and Henrietta painted the name HONEY FITZ on the vessel's prow and Lester and Virginia retaliated by using fertilizer to make the waterfront grass spell out gigantic letters endorsing Nixon.

Virginia also shared a bond with other Hanford wives precisely because they were isolated from their men and from all previous roots. Though their discussions weren't quite equivalent to those which led to the development of new fuel element claddings or remote manipulators, the bridge club sessions did include what Virginia called "talking shop."

"Some of this," she said, sipping a gin and tonic and laughing as she had earlier at her past innocence, "was almost like the commercials where housewives discuss whether or not to use new Lemon Pledge. But it helped break our isolation and taught us how to deal with common problems like what to do when the kids were sick, just as the bridge club also let us use all the polished shiny surfaces we'd cleaned and let us bake cakes and pies to be appreciated."

For all the constraints, women who accepted the basic codes were welcomed by their peers as sisters in the atomic order. They even received commiseration when their men seemed to lust primarily after nuts and bolts. If the husbands were at times socially awkward, tongue-tied and ill at ease, it was the female job to provide the warmth and graciousness lacking in gamma ray monitors and steel-clawed robot manipulators.

The men's distancing and immersion in their mission of invention undeniably created problems for their wives. But Virginia valued what their efforts, at their best, eventually yielded. "Maybe our society does let technology run unchecked," she continued, "by assuming that if people work long and hard enough machines will sooner or later solve all difficulties. But I remember when I first got married and had to clean clothes on a washboard because we couldn't afford an electric wringer. Then we got one and I watched it do by its own power what had made my hands battered and sore and took forever. Maybe your generation missed that feeling of receiving a marvelous gift, because you grew up taking technology for granted, but in the same way that invention has its limits it also has it strengths."

This was true as well of the Richland community, and Virginia remembered going away to see her daughter after her mother died and returning with a feeling of terrible emptiness. "But no sooner did I park my car and start walking to my door, than I ran into four different friends who said they hadn't seen me around and asked where I'd been and how I was doing. That made me feel I lived in a place where people cared for each other and where, sappy as it sounds, we had almost the warmth of a family from knowing each other as long as we had. In that sense all the pioneering really did leave us with something."

Virginia's daughter Allison grew up waterskiing on a course

Lester made in the river with inner tubes, attending Brownie and Girl Scout meetings where they sang songs and worked with crafts, and going out each weekend to the High Spot Club. Once she ran away from home and hid in a neighbor's yard until she got hungry. Another time she joined with kids on the block to make a ketchup-soaked rag dummy which they placed in the street where the AEC head's grouchy wife ended up running over it.

Because Richland was in part just a small rural town, its kids went hiking and swimming and rode their bikes all around the sagebrush terrain. They hung ropes from alphabet-lettered house to alphabet-lettered house, installed pulleys and old tires on the ropes and devised their own wondrous contraptions so they could ride back and forth. They played tin-can telephone, ate out on special occasions at the local pancake house and hated Sunday school no more or less than kids anywhere. At night sometimes they'd drive with their dads to the by-pass highway west of the shelter belt, then stop the car and listen to coyotes howl at the desert moon.

Because Richland was a military base, plunked down in a place with which its builders and inventors had no genuine connection, its sons and daughters suffered to a degree from their isolation. While kids in nearby Yakima might have one relative they called "Airplane Grandma" because she flew in for sporadic visits, Richland children knew none of their uncles, aunts, grandparents and cousins, except across the barrier of 200, 500 or 3,000 miles. Since their parents had displaced the original residents, they were viewed as interlopers by the kids from surrounding farm towns. And perhaps because they were backed by all the power of the atomic enterprise, they in turn shared their parents' attitudes that Kennewick and Pasco were junior league towns, worthwhile at most for periodic shopping trips.

School life had its own hierarchies not always congruent with the way one's dad ranked at the Area. But kids knew who lived in the south end and who in the new and more prosperous northern section (whose managerial occupants and proximity to the Columbia earned it the name "The Gold Coast"). As in any company town, the boys would warn each other teasingly that "his dad's a top boss" when they got into arguments in the playground. When a major strike threatened in 1961, children of union craftspeople and of engineers and managers knew enough about who was on which side to spark major arguments. Even Virginia suggested her boy get together with the son of the AEC head after his mother called and said "I've seen your Steve and he looks like a fine young man. I'd like my David to play with him."

It was convenient that the town had so many young families, because kids would find a dozen playmates their own age on every block. Their friends came from reputable stock, of course, as the town had no "other" side of the tracks and poor people were seen only in magazines or if the kids were taken to Seattle or Portland. Richland's few blacks were, in the words of an engineer's son who grew up here and left, "treated like pets." Although the town had a dozen Jewish families—half scientists and engineers, half doctors, dentists and merchants—Richland's children generally assumed all kids were Christian. With its inhabitants overwhelmingly middle-class, dedicated to their careers, well educated (whether in conventional institutions or through their own efforts) and assimilated, parents considered Richland "an ideal family town."

In a sense it was, because the parents who created this place in a context of wartime fear did everything possible to safeguard their children against all privation, danger and uncertainty. One father prohibited his daughter from watching "Twilight Zone" or "Outer Limits"—not because the shows were trashy, but because, like horror movies, they might lead to nightmares. Although it was safe to play outside even at night, parents were careful always to ask their kids where they were going and who they were seeing. The wives worked as hard at raising healthy, well-adjusted children as their husbands did at making plutonium at the reactors.

As befitting a highly educated community, it was understood that Richland kids should do well in school. Although the sciences predominated and they could take calculus and analytic geometry in their junior year at Columbia High, the school also offered Latin, psychology and accelerated English classes. In addition to spending time with their children in their basement workshops, the Hanford men were happy to help with math and science homework, and many bought their kids copies of the book *How Things Work* to serve as a bible for technical answers. One family institutionalized a nightly geography quiz in which the father would ask "What's the capital of Venezuela?" or "What's the longest river in the world?" and the kids would fire back the answers or run to look them up in the World Book. The pressure lay not so much in being the first and the best, but in getting respectable B's and A's and in learning enough to ride the new advances that would surely be delivered soon by America's inventive technologists.

Columbia High even had a rocket club, sponsored by one of the physics teachers, whose members would build elaborate solid-fueled projectiles and launch them a mile in the air through the clouds that hung high over the desert west of Richland. The rockets had two stages, with mercury switches to trigger the second firing. The kids attached lights, cameras, nose cone parachutes and talked seriously about even including a mouse as a passenger. They built a blockhouse and concrete launch pad to ensure a safe professional operation. The rockets' flaming trails touched off dreams of other frontiers still left to conquer.

To what degree did Hanford's product create a different environment for growing up? The reactors themselves were invisible behind barricades and fences. Kids might draw pictures for their dads to hang on the walls of the windowless buildings where they designed the new systems, but both the work and the offices and labs where it was carried out remained shrouded in a secrecy only incrementally different from that of the wartime era. If your father was one of two hundred or so middle and upper level managers, you postponed going fishing on weekends when, just like a doctor, he was on call for round-the-clock troubleshooting, and you watched him participate in monthly Emergency Evacuation Procedure tests during which the superintendent rang each manager's phone in a continuous buzz until the men picked up the line and waited to respond while their names were called one by one in alphabetical order. If you were home at the right time, you might even catch the FBI cadre walking up to your door, gray-suited G-Men right out of Eliott Ness and "The Untouchables," and asking *your* mom whether she'd noticed her neighbors doing anything strange.

War was generally glamorous for kids growing up in the years following a conflict so heroic that their fathers' sole gravestone markings years later would most often be the names of their units at Anzio, Saipan or Normandy. Everyone knew of the Nike missiles on Rattlesnake, and of Camp Hanford, a military base located between Richland and the Area. The air raid sirens visible all around town were just a sign of wise preparedness. The stories of planes shot down after accidentally flying over the reactors merely fueled adolescent death-thrill lust.

Just like everywhere else, the Hanford kids dove beneath their school desks in surprise blast protection "drop drills." But here teachers often told them that their town's elm- and maple-lined streets were among the Kremlin's top military targets. One engineer's son had a junior high math teacher who showed government films about Hiroshima, Bikini, Eniwetok and White Sands. He talked of the weapons' destructive power, savoring each detail of melted eyes or bodies so vaporized that all that remained of them were shadows burnt into the bridges or streets where they'd been walking. He said there were secret Japanese films so horrible the government would let no one see them. Then he switched his tone and, like a preacher letting the children choose between damnation and salvation, explained the ease of recovering from nuclear attack by brushing fallout off food containers, drinking water from toilet bowls and holing up reading books and magazines until it was safe to come out.

Because only the most visionary built shelters, and because the authorities took seriously the notion of Hanford being specially targeted, the school kids were caravanned in practice evacuations (some went on school buses, some were driven by their mothers) to presumably safe rendezvous points in the desert 20 miles to the south or in the Horse Heaven Hills behind Kennewick. One woman's son woke up crying the night after a class on civil defense and asked her "What about the cow?" and "How do you know you'll find me?" She soothed his fear by assuring him that the real bombings wouldn't happen anyway—but that if they did all problems were planned for and anticipated. Another boy was reassured by his precocious eight-year-old sister that he shouldn't worry abut the Russians seeing the sun reflect off their new yellow Buick, because they could always cover the car with peanut butter "and then eat the peanut butter if we get hungry." Another girl got carsick during the drill, but still thought it an exciting adventure.

As the fathers repeatedly told their kids, preparation made all the difference. What they were working on here would make holocausts such as Hiroshima unrepeatable, and if the children felt anything, it should be a sense of distinction and pride. The Red missiles, in any case, were even more distant and invisible than the complexes of the Area. As one A student pointed out, in order to end a fifth-grade discussion about whether or not Richland was a target, "I don't know what everyone's so worried about when we've got Camp Hanford to protect us."

When the kids reached high school age they attended an institution, Columbia High, whose athletic teams—called The Bombers—wore jerseys and helmets proudly displaying an exploding mushroom cloud. The mushroom cloud emblem, labeled by the principal "a symbol of peace," also went with variations, on a huge green pennant that hung over the gym, on pep club brochures, on bleacher seats and souvenirs, and even on graduation programs and yearbooks. The class of '68 donated an inlay of a finned bomb that was set into the administration building floor. The basketball teams had a three-foot-high green and gold projectile which, until the Pasco rowdies began to cause fights about by stealing, they placed each half-time in center court.

But the team name, even for those who later saw its irony, was just "a cool aggressive symbol"—more modern and more menacing than the animate totems of the Bulldogs, Bears and Braves who competed with the Richlanders in the Yakima valley athletic leagues. It had an aura of power to match the powerhouse basketball teams which went so often to the state tournaments that seniors felt it highly unfair if their year came up and their team failed to make the annual trip to Seattle. Columbia High's emblems were treated as casually as if they'd been images of miniature toothpaste tubes adorning the sweaters of children of Colgate workers.

The mothers as well cultivated a blasé attitude about the cataclysmic weapons their husbands helped produce every day. An editorial in the *Tri-City Herald* suggested, in response to America's atomic tests at Yucca Flats, Nevada, "We have defense mechanisms . . . against enemy atomic attack . . . but because of lack of interest on the part of the public they are dusty and rusty," and concluded that an ounce of prevention was worth a pound of post-attack cure. A civil defense supplement admonished blithely, "If anyone has been outside and fallout particles have collected on his shoes or clothing, they should be brushed off before he enters the shelter again." It compared preparing a nuclear survival plan, "just in case," to the carrying of fire, auto or health insurance.

The women responded, or some of them did, by dutifully stocking basement shelters with canned food, bottled water, blankets, books and games for the children. Some also kept supplies in the trunks of their cars. But for the most part, the wives were far too immersed in running their households, raising their children and creating a civilized cultural milieu to worry whether they were in fact "adequately prepared." Although Hanford was special and they were privileged to aid its mission, the domestic world they oversaw was not really so different from the one they would have faced in any other suburb of the era.

There were a few, however, who considered aspects of polite Richland life close to operational madness. Kendall Ozaroff met her husband, John, in Ottawa, Ontario. He was a physicist working for the wartime Canadian National Research Council. She was the information editor and North American media liaison person for the Free French resistance. When Ozaroff saw her sitting in a restaurant holding a copy of *War and Peace* and wearing her career-girl brogues and tweed, he asked, "Have you ever thought of reading that in the original?" She explained that Russian wasn't one of the seven languages she knew. They began talking about literature, art and music. Soon afterward they married.

After the Canadian sector of the Manhattan Project was established at Chalk River, Ontario, Kendall and John went up to join it. As in the early Hanford days, people here were intensely dedicated to their mission. Kendall never knew exactly what her husband was doing; she even trained herself to shut her ears when he talked in his sleep. But as she'd left her own writing work, she did her part now by keeping their apartment in perfect order for him to come home to through the subzero winter weather, by taking all responsibility for their two young boys and by not even asking him to do so much as take out the garbage or anything else that might distract him. She felt part of an urgent historic project.

The Ozaroffs came to Hanford in 1948, following his completion of a war-interrupted doctorate at M.I.T. Because John was a top theoretical physicist, they moved into a house overlooking the Columbia on a street inhabited by the other Richland higher-ups. He spent time with other physicists and with a few of the chemical engineers (they looked down on the AEC administrators because "all they create is rules and regulations"), built his own hi-fi system, and listened to Chopin, Tchaikovsky, Stravinsky and Beethoven on it each evening. But Kendall began resenting the endless rounds of social teas, bridge clubs, theater clubs and garden clubs: the women who didn't even like children but started basement nursery schools because they were bored, who wore black moiré dresses to hospital auxiliary meetings, and who couldn't talk of anything except what new furniture their husbands had just bought or how well their kids were doing in school and Little League. She'd still pour Ozaroff's Scotch when he came home and supervise the kids' baths and get them ready for bed; they'd still sit in the garden and talk about literature; she'd still listen, envisioning herself as a mouse in the corner, while his friends and he discussed the nonclassified aspects of their work. But in a manner that presaged the shifts between this, the General Electric era, and a later one of inflated salaries and massive cynicism, Kendall lost the sense of urgency and shared participation she'd once had. With her kids now off each day in school, and with her writing having been shelved indefinitely when she became a housewife, she decided to resume her girlhood pastime of riding.

The horse she bought—a black mare named Thunder—was strong and spirited. After getting up each weekday morning to send the kids off to school and her husband to work, Kendall drove to the stables, just before the bridge across the Yakima that led to West Richland. Loading up saddle panniers with extra food and water in case anything happened to either of them, and wearing a cowboy hat and kerchief to keep the sun off her head, she rode for seven hours a day—out into the desert sagebrush, along the Yakima or Columbia, or toward Kennewick. The rides weren't particularly a time for daydreaming, Kendall being a practical person who checked constantly where she was going, whether the horse was thirsty, whether rain or dust storms seemed likely. But she could head across the sand in whatever direction and as fast or slow as she wanted. She could get back to "the fundamentals of the sky and earth and clouds." She could feel "a kind of freedom which unless you've ever ridden like that you just can't know."

Kendall spent seven years at Hanford and rode nearly every day while her husband was working. But John stopped discussing Tolstoy and Picasso with her, talking only "about the price of cars he was looking at or how so-and-so got a raise or was promoted." They didn't share the music any more, "he'd just go off into it like a miser sitting on a beautiful painting." Now, when she said she was thinking of taking up Russian he discouraged her. She persisted. He reiterated: "I don't think you should." And on one of their regular trips to see his parents in British Columbia, he stopped their yellow Mercury convertible by the side of the road. She thought it was "so the kids and dogs could get out and piddle," but without saying anything John reached behind the seat, picked up about a dozen books he'd put there, and began throwing them into a river below.

"What the hell are you doing?" Kendall asked.

"These are Russian books," he said. And while the kids and dogs looked on, while she cursed and called him an idiot, he explained that this was no time to be learning that language. She kept silent the rest of the trip. She knew "you never put a person working with atomic weapons on the spot." But even though he had to have had his expedient reasons, his action had made her feel so dependent and helpless that she realized a place that made him do that could never again be her home.

Because security prohibited repeating concrete details and because workers kept any general doubts to themselves, Kendall never heard of plant mishaps. But she began to have misgivings about nuclear safety and they came to a head shortly after the incident with her Russian books. It was 1953 and the government was holding a Nevada test for a bomb whose plutonium had originally been milled here. Because the fallout was scheduled to blow over Hanford, John mentioned it to her in passing. She was riding her horse, one day as always, when she noticed his car parked, along with those of other Area managers, in front of someone's house. She thought about the household world she'd worked so hard to put together—and about how it was about to be invaded by a menace in part created here. Then she went in—she was the only woman in the room—and asked: "Do you gentlemen care at all that you've caused this and that now it's affecting us, your wives and children?" After being told "We have this well under control," she left.

When Ozaroff came home she expected him to be mad, but he acted—as he always did—as if nothing had happened. When she asked, the next day, if the danger bothered him, he answered only, "Well, I guess you should stay inside."

They divorced soon afterward and Kendall moved to the nearby city of Walla Walla to work as a journalist.

3
Pioneer Life

Like the Manhattan Project migrants, Hanford's original settlers hoped to pioneer a new world. In a 1910 brochure issued by Hanford Irrigation and Power Lands, peach trees were pictured near collapse from the weight of their fruit, maps showed irrigation trenches running as broad as small rivers and charts explained a dozen ways to get rich off the soil, climate and newly available irrigation water. "The man who owns his own fruit ranch in this valley," the brochure promised, "is one of the most independent men in the world. Ten acres will yield him the salary of the average bank president. He is dependent on no man: his harvest is always certain and profitable. . . ."

The new migrants read these words and others like them, and arrived, driving in on dirt roads from Yakima and riding the stern-wheelers that came up the Columbia or the Northern Pacific trains that ran through Pasco—then found the irrigation trenches leaked so badly that even after hand-carrying water from the river, many lost half their crops during the early years.

The farmers eventually took over the water system, naming it the Priest Rapids Irrigation District after a dam just upstream. Though they were still subject to the cons of market middlemen and the caprices of the land they worked, they raised their peaches, apricots, cherries and plums, their apples, potatoes, wheat and alfalfa. Where necessary, they still hauled water by hand and set out smudge pots during the winters so their orchards wouldn't die. They swam in the Columbia and carved wagon ruts to lead sheep across it one winter when it froze. They traded with the local Indians (one kid, himself one-eighth Cherokee, tried unsuccessfully to barter an old pistol for a papoose he thought was cute). They danced, drank at White Bluffs' Oasis Saloon and built the Hanford and White Bluffs schools. When the war came they assumed the isolation of this land they'd made their home would ensure their safety.

But the orders came to leave and the farmers received $300, $400 or $500 for land which, together with its improvements, had cost them far more than that twenty years before (some sued after the war and were paid slightly more). When they tried to resettle elsewhere in the region, landowners jacked up prices two- or threefold to exploit the shortage. They were left on their own to reconstruct the homes and the community their dispossession had shattered.

The settlers' Hanford, for all its problems, had been as warm and rich an environment as any agricultural area. Atomic Hanford, in contrast, was a barren world—on the order of all the relentlessly developing industrial enterprises which, as historian Lewis Mumford points out, took their model—because it was the first entirely inorganic environment to be created, and worked in—from the mine. The mine, Mumford says in his book *Technics and Civilization,* is an environment where in "hacking and digging the contents of the earth, the miner has no eye for the forms of things," and sees only sheer matter that it is his duty to break through stubbornly. It is an "environment of work: dogged, unremitting, constant work . . . a dark, a colorless, a tasteless, as well as a shapeless world: the leaden landscape of a perpetual winter."

This description of gray abstraction and distancing could apply—cosmetic decoration aside—to nearly any modern factory, computer room or office. But in the Area, with its windowless monolithic buildings, it speaks of an environment where, because all surroundings are man-made, those left untouched are considered as useless as the matter a miner breaches, then discards in his search for gold. What one top Hanford manager called "the new post-industrial future" would be a triumph over a capricious natural world. The interchangeability of the ore that the miner produced would be extended through an array of interchangeable megatons for unlimited security and kilowatts sufficient to fuel an infinite consumption economy. Unchallenged by what Mumford would later call the "variety, individuality and diversity" present in all natural life, the Hanford culture would find its challenges in creating machines to meet all manner of human needs.

For the old hands who lived their lives in a continuum between the Area workshops and those in their basements and garages, their inventive creations constituted their own language. They believed in what they could see, prove, make or alter with their own hands. They viewed their external surroundings—the adjacent desert—as so grand that no one could harm them and so virgin that one could create any number of miracles here without disturbing previous human efforts. Equating mechanical efficiency with security and happiness, the old hands used their skill and knowledge to build machines that would deliver more and more plutonium product. If one questioned their inventions, they assumed all apprehensions were due to technical ignorance.

The military and corporate model from which Hanford developed pushes all attitudes and actions toward approaches that can best be termed "instrumental." If a resource or a relationship is useful for a particular goal, it is valued. If it exists autonomously, in and for itself, it is at best irrelevant and at worst an obstacle to be destroyed or altered. Whatever are considered higher purposes, like those of the Area's mission, take precedence over the lives of ordinary individuals like Hanford's original settlers or the islanders forcibly removed from the Pacific atolls where we tested our H-bombs. Even those who believe in the overriding goals must subordinate any impulses to question fundamental directions or even—as in the case of Hanford's workers—to discuss what they are doing with people not certified participants in the "cause." As they strive constantly to create more efficient machines, they waive judgment on what would occur should their plutonium product ever be used.

One explanation of the old hands' pragmatic and mission-oriented approach came from Sam Beerman, Virginia's former next-door neighbor. Sam came from Seattle in 1948 to work as a chemical engineer at the old reactors, switched to the N plant in 1963 and five years later started his own job shop located in a small edge-of-the-Area think tank row resembling a miniature version of Boston's Route 128 or California's Silicon Valley. We met at Sam's new North Richland house which, with its beveled glass Chinese-handled doors, dark sloping roof and Ford LTD in the driveway, could easily have been perched in the Hollywood Hills of LA. Sam is Jewish, carrying the sort of kind, sad expression that might well grace a Saul Bellow character, wearing the bermudas, corduroy slip-ons and polo shirt of a suburban dentist or merchant, and speaking with inflections and rhythms which contained a stop-reflect-and-hesitate air I knew well. When I asked how it felt to live in this churchgoing Christian culture, Sam said he'd never felt any anti-Semitism, although he did have friends here who had.

Sam had come here assuming he'd eventually move back to Seattle, but Hanford gradually became a permanent home through a combination of career promotion and encouragement, the building of friendships with his co-workers and neighbors, and the erosion of old family ties as his wife's parents died and his moved away from their Seattle home. When I asked if Hanford's practice alerts made him fear nuclear attack, he said they were "just wise precautions" for either war or internal accident. Though they made sense "in the same way that a fire drill does," he'd always believed Hanford was too early in the weapons creation chain to be a first strike target.

I had talked the day before with the N reactor operations director and asked him how it felt to work on the Nagasaki bomb. "You have to realize," he explained at first, "that most of us had little idea what we were creating. Nagasaki did do the trick, because they kept fighting till then, but I don't like making atomic weapons. I think the more you have the more risk of nuclear war. That they can even conceptualize a neutron bomb completely terrifies me."

He sounded almost like a pacifist. When I asked "Does that mean you favor disarmament?" his head jerked back violently.

"That's my personal business," he replied, shifting forward flushed. "I touched the Liberty Bell. I consider myself a patriot. I'd like this to stay the greatest country in the world. Since we do have enemies, we have to leave judgments to the military leaders, not become weakened and run over by ruffians. I don't favor unilateral disarmament, and if the government experts—who know more than I do—feel we need a neutron bomb, I have to back them up."

When I recounted the director's story to Sam, he said, "That position makes sense to me because what one wants and what one supports aren't always the same. It's that man's job to run a nuclear plant, not decide policy, and I also—once the military made a decision—wouldn't second-guess it. I think the problem," he continued, "is that people have gone from reacting against institutional authority to mistrusting authority of knowledge. If Nixon told me to jump in a lake I'd say 'Go to hell,' but if a doctor said the same thing I'd follow his instructions." Now, though, there were so many experts on different sides it was hard for people to make up their minds. But Sam knew that those in charge were friends and neighbors whom he trusted. Richlanders were a self-screened group to begin with. He thought that, "although not blindly," his community still believed "in carrying out orders and respecting authority."

Sam's fifteen-year-old son hunched over on the grass, painting a set of water skis and asking constantly, "Is this right?" and "What should I do next?" His wife, Henrietta, came out briefly to warn the kid, "Watch out or you'll get orange all over the lawn and on your shoes," and asked if I was "one of those Three Mile Islanders."

When I said I wasn't, she asked, "Well, are you part of the group that came with Ralph Nader?" And after I said I wasn't that either, she explained, "I don't like groups" and went inside to fix us juice.

Sam continued, straining to go beyond simply telling stories to explain the roots of what he and the other old hands believed. He admitted, a bit sadly, that situations such as Vietnam did test the wisdom of obedience. He said "we were being lied to there," he guessed Richlanders did oppose the war less than people elsewhere, and he agreed few here ever had basic qualms about the nuclear industry. "But in a democracy we're given the vote as a means to shape decision-making. When we elect somebody it's our responsibility to implement their programs.

"Maybe we were products of a more structured and less secure time," Sam continued, stopping as he talked to reflect on each phrase and sentence. He recalled the Depression. Although his father kept his job at a clothing store and although the family never really went hungry, they scraped "pretty close to the edge" and the vacant lot next door was always full of thin old men and women trying to gather edible dandelions. The difference between growing up with that experience and with the comfort he'd tried to give his kids had to have some effect, "because when we keep repeating stories about how we wore pants and shoes handed down from five brothers or when we lecture 'Now clean your dinner plate because you know there are starving children in China,' it's at least partly because the fears from that time will always be with us.

"You have to remember that after we pulled out of the Depression and faced and won a terrible war, it's understandable we'd think of Roosevelt as our commander-in-chief, that we prized as something special a feeling of esprit from having all pulled together," and that Americans would resolve to ensure the economy boomed forever and the military always triumphed. Sam's daughters, in contrast, were "products of a less structured, less embattled era," and after he raised them in "this good, safe family town," they ended up marching in Seattle's Kent State demonstrations.

"As far as how I've raised my kids," Sam said, my own mother says I'm too softhearted. But I want them to think." He pointed to his son, who looked up startled after asking for the tenth time, "Now what should I paint next?" "I want *him* to think and make the decision on the damn skis, not me."

Sam had visited England's Windscale plant a few months after a reactor caught fire and contaminated the countryside, and saw everything back to normal with cows grazing and being milked, vegetables growing and people working in their fields. He knew Canada had repaired its Chalk River complex after a major near-disaster there involving a burning fuel element, and that people now lived normal lives in Hiroshima and Nagasaki. "It's not that dropping those bombs wasn't terrible," he explained. "But given that it was as awful as anything which could happen, I think the talk about nuclear accidents making places permanently uninhabitable is overstated." If one thought about natural disasters such as those produced by Krakatoa or Hurricane Betsy or even some tornadoes, what man could do was "pretty puny in comparison." Sam believed nuclear power was "a manageable technology," and that when problems did occur they were correctable.

"Maybe," he concluded, "we should regard running these reactors as a calculated risk, like that of driving an automobile ten thousand miles a year. What we produce here may have developed from the bomb, but that was a tremendous undertaking. Now that we have that breakthrough advantage, for God's sake let's run with it."

Since Hanford's original residents were evicted at the project's start, the atomic pioneers faced no previously existing traditions. Because extended families were left back home no grandparents were available to offer their perspectives from an older, less technologically dependent time. Although the new migrants came from a variety of regions and even different ethnic backgrounds, this original diversity simply made them strive twice as hard to create not only a common language of accents and phrasings, but common outlooks, ways of thinking and ways of being. As Sam explained, the drive to blend seamlessly into what could be called an assimilated generation was the product not only of a specific technological base, but of an era when celebrating American power became a given throughout the nation.

Virginia viewed this celebration as rooted in Depression-era vulnerability which those working or coming of age during that period had resolved never to repeat. "If you didn't elbow someone out of the way," she said, "you wouldn't get a job no matter how brave, kind, generous or gracious you were. If you didn't get a job you couldn't feed your family. While your sense of equity and fairness didn't disappear, it became a luxury to indulge in only after you'd found secure work. Of course, there still was an 'I'll help you and you'll help me' feeling within family units—you'd have been sunk without it—and maybe it extended to your ethnic group or immediate neighborhood. But there was always a line where the sharing stopped, where you ended up out for yourself and your immediate circle. I think our desire never again to be powerless and to make sure we could get the other guy before he got us had a lot to do not only with Pearl Harbor, but also with the economic hardship just before."

It's possible as well that this flight to conventionality was rooted in World War II's unprecedented horror. Both Auschwitz and Hiroshima, as Robert Jay Lifton has pointed out, destroyed not only their victims' lives, but also all contexts from which they—and those who survived these holocausts—could draw meaning, individuation and humanness. If the old hands and others of their generation readily accepted a society run by giant corporate and governmental institutions, perhaps their actions were rooted in part in the notion that a basic natural order had now been shattered.

In any case Hanford developed in an era when Americans all across the country returned from the fighting and moved into suburban garden cities secured by government loans and connected by endless new highways. They left behind, as did the atomic pioneers, the extended networks of relatives and ethnic kin that had supported them in immigrant days. The smells of Sara Lee and Swanson's replaced those of home-cooked lasagna and lutefisk, old country history was no longer even mentioned and the old languages no longer taught. The trimming of hedges and manicuring of lawns replaced agricultural labors. Dissent became buried in the National Labor Relations Board institutionalization of the union movement and the post-Stalin backlash and McCarthy-era suppression of the left. The relics of preindustrial culture were given final burial by television screens showing only a nostalgia-laden past or clean domesticated present.

In some ways the sensibility of Hanford's founders resembled that of the healthy white Westerners Wallace Stegner described in a 1964 essay entitled "Born a Square." Homesteading as an imported population on empty terrain, the people he spoke of were "products of a world still nascent, and therefore hopeful." Their character embodied a "naïveté of strenuousness, pragmatism, meliorism, optimism and the stiff upper lip." Because their recently settled territory contained "the only American society still malleable enough to be formed," it symbolized "the New World's last chance to be something better." The Richlanders viewed themselves in these frontier terms, and like Stegner's pioneers and Hanford's original farmers, their labor did make a new land their own.

But if the Richlanders lived, like the Westerners, in an environment far from ghetto poverty, urban angst and overt racial tension, they labored not in forestry, farming or ranching, but in an industry whose potential effect reached all across the planet. The vistas stirring Hanford workers' dreams were not those of the terrain they settled, but of inventions—converted from their military origins—they hoped would someday create an energy cornucopia for the world, and of the comfortable respectable home lives they were able to build for their families. If the naïveté of Stegner's squares stemmed from lives in which everything was possible because their culture was barely created, that of the Hanford pioneers came from importing mores and attitudes FBI-certified to be friendly, folksy and compliant. Hanford workers were pioneer builders, but they labored to serve the needs of a mammoth military industrial enterprise.

Back at Virginia's, I met Ed Wyckoff, the forty-year-old son of one of her friends. Home on a visit, wearing jeans, a black turtleneck and a neat beard, Ed asked his parents, "You mean you came here not knowing this was a nuclear plant? You could have endangered your lives."

His father argued briefly, explaining "secrecy was necessary" and insisting, as if it were empirical proof, "It was safe. We're all here together now."

"But there were accidents. At least you could have been told."

Ed's father tightened his lips, looked off and began talking about golf with one of the other men in the room. His mother mumbled a vague agreement. A friend of hers from the Richland Players fiddled with the elegant eyeglasses she hung on a chain. Ed talked further of growing up and leaving for New York "believing that no crime or prejudice existed anywhere"; of getting a car at age sixteen and driving the empty highways "just to go somewhere."

"Then I read Joan Didion's *Play It As It Lays*," he said, "and I thought 'that's why I'm racing, because if I stayed in one place here I'd go crazy.' "

PART II

MODERN TIMES

"At least I'll check the welds more carefully than someone who doesn't give a shit . . ."

—Hanford pipe fitter

4
Second Generation

There's a story going around Hanford about a visiting Easterner who asks a cowboy to explain his dress. "Well, this hat," the cowboy says, "shields me from the sun. I can use it to cool myself, and if I come to a creek I'll drink from it. This kerchief keeps the dust out and mops my face. The shirt snaps so I can take it on and off in a hurry. The jeans are tight and rugged to hold up well and not give me saddle blisters."

"Great," says the traveler, "but why are you wearing tennis shoes?"

"These," replies the cowboy, "are so I won't be mistook for a pipe fitter."

Though only half the pipe fitters, ironworkers, carpenters and other craftsmen building Hanford's three newest reactors for the WPPSS utility consortium dress in their free time like Texas range hands, nearly all are roamers who follow their work from place to place like Chisholm Trail drifters. Their jobs here are little different from ones previously held constructing Louisiana oil rigs, Idaho Nike bases and the Alaska Pipeline. There will always be other projects elsewhere.

Although Steve Gire worked not for WPPSS but as a consultant for the computers that Westinghouse used for its breeder project, he did his own roaming, spending between $3,000 and $4,000 a year on bi-weekly trips to Seattle and Portland. Gire, now forty-eight, came here from Cincinnati in 1959, and because his father was a blacklisted Communist Party machinist, had to "become like Woody Allen in *The Front*," and take a loyalty test before working. Gire was "enough of an Uncle Tom" that he got a job researching designs for nuclear-powered aircraft. He went from that project to years working on fuel fabrication. He now made $30,000 a year developing software for safety systems, shielding requirements and accounting procedures.

I visited Steve at the low ranch house on the outskirts of Kennewick that he bought for $14,000 in 1965, and that was now worth well over $60,000—but "would drop to floor-level price in a month if they ever stopped new atomic construction." Six-foot-three, wearing mirror sunglasses, sporting a full beard and white hair which hung halfway down his back, he looked like a cross between Santa Claus and "Bonanza" 's Ben Cartwright. He had changed from his work slacks and shirt into a pair of red swim trunks. He stir-fried tomatoes and hamburger meat to make burritos. The Grateful Dead—playing from a live bootleg tape he picked out from over a thousand cassettes and records on a shelf—traced the slow lazy tendrils of "China Cat Sunflower." Since Steve didn't have air conditioning—not due to the cost but because "I just like the climate"—the desert heat rippled in, cooled slightly by a swimming pool at the rear of the house.

Echoing Ed Wyckoff's explanation of the need for constant motion, Steve explained, "It's impossible for anyone sane to stay here every day without flipping." He reminisced about San Francisco, and visiting the early 1960s North Beach scene which included people like Lawrence Ferlinghetti and Paul Krassner. He said he missed the hippies. He thought the only worthwhile culture now was music, and we talked of Patti Smith and of a Who concert he made a special trip to New York just to see.

"The people running this place," Steve continued, changing the subject, "admit nothing bad exists. When a rapist at Columbia Basin College attacked five women in a couple weeks' time, the *Tricycle* [*Tricycle Herald* being the name among the cynical for the *Tri-City Herald*] didn't even mention it until people found out by word of mouth and began to make a stink." The DOE might now funnel its money through a dozen corporations instead of one, as in the old General Electric days. But if Steve was blackballed by one, he would bet none of the others would hire him, "and that's a company town any way you name it."

I leaned back on a turtle-shaped floor cushion as Steve sprawled regally, belly hanging over his trunks, on a beanbag chair. When I asked if the cover-ups made him worry about the plants, he joked about being "just another of your base-line pragmatists figuring that because I haven't died yet, I'm not about to."

Steve lamented again the lack of culture here. Yes, he preferred smaller towns to overloaded metropolises, yes he liked the feel of the desert, and yes the wage scale was better than any he could earn elsewhere. But Tri-Cities was so new and so much a single industry creation that all life here was subordinated to Hanford's production rhythms.

He described a book he'd read recently, in which Julian Jaynes used Homer's *Odyssey* to explore how our rational and manipulative left brain hemisphere "began taking over more and more until we end up with places like Hanford and a human race resembling one of those companies that goes on, even though everyone knows it's dead, past its time."

We talked on about James Joyce, *A Clockwork Orange*, and Ivan Illich. But the speculations and arguments were only disconnected words, meaning nothing to Steve beyond a chance to strut.

As the Dead sang "The grass ain't greener, the wine ain't sweeter, either side of the hill," Steve began talking again about Hanford complacency. "You've heard of Steve Stalos, the scientist who quit his job here?" he asked, and explained that—until Stalos gave descriptive documents to the Seattle papers—no one even knew about a 1954 gas leak which deposited radioactive material on a 400-mile trail between Portland and Spokane. "Coal miners at least realize they're risking their lives for high pay. But people here treat the reactors just like the erector sets they used to build as kids. If they weren't all so busy playing house, they might notice this place is not just supremely ugly, but also a lot less safe than they think."

I thought of Jaynes's reference to the Odysseus myth, and of how the German philosophers Max Horkheimer and Theodor Adorno explored the same story in their book, *Dialectic of Enlightenment:* The sailors' ears plugged with wax so the Sirens' ballads wouldn't tempt them foreshadowed modern workplaces where you didn't know what you were creating and where talking, laughing and singing were prohibited as distractions from the job. Odysseus, the commander, was able to listen, of course. But tied to the mast, he became the first modern bureaucrat—knowing and watching all, but subordinating everything and everyone to the mission he had to complete.

The same story applied, of course, to those who commissioned the bombs and ordered them dropped. It applied perhaps even more closely to those who maintained fragmented vision as they climbed, tiny fly specks, so they could weld and wire together the massive grids of the reactor scaffolding, and to those who sat inside sealed buildings to run machines that ran other machines that built, tested and harnessed a web of paper memos, lead-shielded windows, $1.29 circuits and $185 million containment buildings—all surrounding and controlling the rods of processed uranium. But Steve knew he was no grunt moving with wax-plugged ears to songs he couldn't comprehend. He took solace, not in believing he was producing global energy, but in using the money and time his job provided to live the real life he cared about in places where he had neither commitments nor stakes. He was telling me about the best unknown Seattle restaurants when the door opened and a young friend of his walked in.

Jim, who worked with Steve at the breeder project, traveled to Portland or Seattle "almost twice a month, ever since I learned to drive." He slid by the 250 miles with the aid of a little beer, a little dope and, when in a high rolling mood, Canadian whiskey; making the trip in three and a half hours because, as Steve said, "Who's going to waste their time going

55?"; not worrying about whether Hanford will mix with the city worlds because the two were eternally separate.

Jim talked further about how Kennewick people were "just conservative rednecks," the Richlanders "just liberal rednecks," and how he disliked the cone of silence which descended, "just like in the old Maxwell Smart TV show, whenever your criticisms go beyond routine gripes." But he had great job security. He liked being around a familiar environment just as Steve did. He could always get away on the weekends.

Jim told Steve about the time in school he watched a scientist hooking tubes up to pigs in blood-transfer experiments. "When I asked if the pig liked it, the guy said he wasn't sure, but guessed the animal would rather be out in a field chomping corncobs. The next day I felt a little stupid, so I explained that I didn't mean to be a gushy humanitarian. The guy said, 'Well, I guess he didn't like it; he died this morning.' "

"The pig was doing it for a very good purpose," replied Steve. "He got paid."

If Steve's affluence saluted a mobile boom economy, Richland's old hands justified their considerably less hyperpaced prosperity by the Calvinist effort that created it and the proper family virtues it served. Yet the period that ran from the mid-1950s through the late 1960s ended by offering the Hanford workers a sense of well-being they'd never previously achieved. Because the tinkerers now owned their own homes, they built new porches, fireplaces, garages, patios and extra bedrooms—and decorated their facades with trellises and wrought iron railings, picture windows, colonial columns and brown shutters with the curlicue carvings of Swiss chalets. A few two-story houses—such as the A model duplexes now occupied by single-family owners—even had their second floors knocked off to convert them into more up-to-date ranch styles. The old coal-burning furnaces were replaced by modern electric heating. The teenagers who would soon leave home and perhaps end up working their own Area jobs and raising their own families could now gather with their friends in remodeled basement rec rooms.

The Richlanders had built a town symbolizing the American good life. The AEC helped, of course, by continuing to subsidize the municipal budget. Wages were sufficiently high so Hanford workers could afford new cars, boats and campers. The isolation eased with a trans-Hanford highway that wound, on the reservation perimeter, past a giant sign announcing that red lights would flash when radiation danger was present. (The sign, which many people felt was too menacing, was finally taken down in the late 1970s.) And on April 8, 1961, *Look* magazine and the National Municipal League honored Richland's newly won independence by naming it an All-American City.

At the plants, the original tinkerers had advanced by now to positions of substantial responsibility. Clark Reitnauer received seven promotions between 1955 and 1962. Sam Beerman and John Rector made equivalent gains in status and salary. Having come here without even knowing their product, these men and others of their era now headed teams developing new loading systems and fuel claddings and separations processes. It was true that with the wartime crisis ended, they now had to put up with the AEC bureaucrats who—like all government workers in the tinkerers' opinions—had license to harass and hamstring the men who got the "real" work done. But the AEC rules necessitated only minor griping and a few subtle evasions. The need for systems to work perfectly, from the time of installation, was one more challenge—solved by building mock-up reactors in the labs and using them to test all new components in every fashion possible. Each invention brought the Area complexes to succeedingly higher levels of efficiency.

While this steady progress was occurring, construction workers spent from 1959 to 1963 building a ninth reactor—N plant—which would produce both plutonium and electrical power. While the initial eight plants were cooled by water from the Columbia running through them, N had a closed loop system, recirculating water through the core and converting it into steam. The steam turned the turbine generators of the adjacent Hanford Generating Project, built between 1963 and 1966 (John F. Kennedy presided over the facility's groundbreaking). When it was completed, the plant's 860-megawatt power output was the largest of any reactor in the world.

I saw Sam Beerman next at the Richland Kiwanis club, sitting at a Hanford House banquet table along with Joel Wyckoff, an engineer named Bob who'd built a hovercraft in his garage workshop and three other Area old hands. We ate roast beef, mashed potatoes, peas and lime Jell-O, sang "My Country 'Tis of Thee" and "Oh Canada" and recited the Pledge of Allegiance. A young real estate broker donated three dollars to the charity fund in honor of the Bomber football team's upcoming state play-off game. Others contributed to celebrate various anniversaries, weddings, and job promotions, and for violations of the half-joking prohibition against advertising one's personal business or cause on Kiwanian time. A card was passed around for a friend who had cancer.

The men I sat with talked sports, reminisced about projects they once worked on together and exchanged gripes about Federal Building bureaucrats. More boisterous and less familial than the bridge club women, they nonetheless shared a similar affection based on thirty-five years of comradeship. At the speaker's podium, Reverend Joe Harding of the Central United Protestant Church introduced the minister who was going to give the day's talk, a friendly rival from another congregation named Willy Rees. "We all appreciate Willy's outfit," Harding said, referring to Rees's brilliant green sport coat and loud fuchsia tie, and proceeded to recall when Rees was teaching a class "and said they could get out of a test by answering one question. 'If a cockroach takes thirty minutes to kick half a hole in a pickle and angels take ten minutes to brush their teeth, how old am I?' They all gave up," Joe recounted in the tones of a seasoned showman, "until a lanky guy in the back raised his hand and said, 'Forty-four.' 'That's exactly right,' said Willy, who of course was a little younger in those days. And Willy looked at him and asked how he knew. 'Well, I have a brother who's twenty-two and half nuts,' said the kid, 'and you're completely crazy.'" Harding paused as everyone laughed and clapped, then concluded with "Ladies and Gentlemen, Willy Rees."

Like a sax player in a cutting contest, Rees began his own teasing one-upmanship by acknowledging the introduction "from my fine colleague Joe Harding." Since Thanksgiving was almost here, he praised America's special heritage, and continued about sharing and love, fellowship, friendship and positive feeling. "Do you ever read those letters to the editor which are always griping or complaining or setting someone straight?" he asked. "They're like the parable of the man who wanted eggs 'one fried and one scrambled,' and when his wife made them without even complaining, he threw his hands up in exasperation, saying 'You fried the wrong one.' Maybe things would be better if we looked more often on the bright side."

The Kiwanians stood and clapped as they had a dozen other times for a dozen other people. The meeting ended and I stayed around to talk with Sam Beerman.

"My friend Roger calls Kiwanis just a bunch of old Republicans singing songs," Sam said, while a bored young waiter rolled up the American flag that had been displayed for the meeting. "But for me it's a good chance to have lunch with friends and visit."

When I asked Sam if he felt people here were too quiescent and accepting, he answered, "I don't think so, because after a person casts their vote, which is really the only thing they can do to change policy, they have to unite with everyone else to get things done. They can work to try and replace the decision-makers, but there's a responsibility to carry out orders. If there's an opposition, maybe it should be, as the British call it, a 'loyal opposition.' I think that's a generalized feeling among people of my generation, and I know it is among all of us here."

"And among your children?" I asked.

"I remember in the late 60s," Sam said, "when my girls were both in college and I was visiting them in Seattle. We saw some kids in dirty sweatshirts and ripped apart jeans, and when I said I didn't understand why they chose to dress that way, they told me: 'You see, Dad, it helps level people and make everyone more equal.' In my day only poor kids wore jeans and we had little enough already without emphasizing it. It confused me that they'd try to look like outsiders.

"It's a danger to generalize, but it seems the desire to excel and stand out from the crowd became less pronounced in kids of that time than among those who grew up in my generation, twenty or thirty years before. Just to use one example, none of them even thought that if Harry Truman, a drugstore owner's son, could grow up and become president, so could they. Instead they wanted to be part of a rebellious crowd, even a mob. They wanted to buck everything, change the country from the outside without going through the means of change that we already have. It didn't work because Americans are too comfortable for revolution; so finally they got smart and became active within the system."

Remembering that Sam's daughters had been involved in the Kent State marches in Seattle, I asked how they now regarded Hanford. "I imagine they might still question things more than I would," he said. "It's true they aren't involved in the nuclear industry and the fact that we've never argued about it may be due only to some kind of respect for me. But I think that, like most kids raised here, they see atomic power as something familiar because they grew up with it, something which—if it's used for human good—they have no reason to fear."

But was Sam correct about the impact of Hanford on its children and what it represented to them? Robert Jay Lifton has said the atomic bomb affected the entire post-Hiroshima generation by taking a normal childhood struggle "to understand death and come to terms with its inevitability and finality," and interposing images of massive grotesque holocausts that have led death to be viewed not as a natural process but as annihilation. Given that most Hanford kids knew nothing more of Nagasaki and Hiroshima than did their compatriots in Detroit, Omaha or Brooklyn; given that their fathers remained tight-lipped about their work and their mothers worried less about waste leaks or atomic blasts than about correctly waxing their kitchen floors before the next bridge club meeting; given that Richland culture prided itself as much on its wholesome, all-American unexceptionality as on the world-altering processes its workers shepherded: Did Atomic City's children develop at least an edge on the holocaust awareness Lifton sees hanging, like a distant shadow, over all growing up in their era, or did they replicate their parents' pragmatic tractability?

Though Rick Price's father, John, never told him how he helped design the milling equipment used to manufacture the plutonium H-bomb hemispheres, or the special racks—called "Christmas trees"—which separated them to prevent an accidental reaction, Rick knew that the "machinery" John said he worked with was awfully important. Born in Richland in 1953, Rick moved the next year from a wartime A house to a home his father hand-built in a subdivided Kennewick peach orchard. As Rick grew older and showed an interest in science, John encouraged him by taking him outside to watch Sputnik pass, holding him as a terrified eight-year-old over the edge of Grand Coulee Dam so he could see its mechanisms and suggesting at any sign of technical curiosity that he too could be an engineer when he grew up. But John, like all the other old hands, would never discuss his work with his own son.

One afternoon, at age eleven, Rick heard a radio song about the scare over fallout-polluted milk. Its chorus went "Strontium 90—there's plenty enough to go around," and while his father said it was "stupid nonsense," it unnerved Rick that the song might be referring to Hanford's product. Later a teacher said he should never melt and drink snow because particles from Red Chinese bomb tests might have polluted it. His parents said that too was ridiculous, and Rick continued being a normal kid, thinking little more about bombs and radiation than he would have growing up anywhere else.

When Rick was in high school, John built another house farther out on the desert edge of Kennewick, with a $4,000 backyard workshop three times as big as the house. Rick's brother Mike called the shop "the model garage" after a *Popular Mechanics* columnist who called himself "Gus of the model garage," and John spent his spare time there happily tinkering with the guns, machines and antique cars that people brought for him to repair.

Rick and Mike, in the meantime, began feeling uncomfortable regarding their role as interlopers in the region. Wanting to be accepted rather than resented by the original Kennewick residents, they strained to touch pre-Hanford history by talking with old farmers, scavenging for artifacts in abandoned buildings and exploring the surrounding desert, hills and marshes. They collected stamps, coins and comics, guns, books and magazines, signs and tools and anything else that would give them a past going back farther than the Area's twenty-five years.

By the time Rick graduated from Kennewick High School in 1971, Americans in most places had been challenged by antiwar dissent. But Hanford's isolation and military-based economy kept Tri-Cities streets largely quiet. When American troops were first sent to Vietnam, most old hands supported the effort. Our involvement wasn't a major subject of discussion; the Hanford men and women preferred, as always, to operate their reactors or raise their kids, and to leave political controversies to others. When they did express feelings, they coincided with the Area's generally pragmatic, get-the-job-done attitude.

At the beginning Vietnam seemed just a routine police operation, cleaning up the politics of one more third-rate country. But when the fighting dragged on endlessly and our government seemed to be withholding its full military punch, the old hands grew discontented. "If we were going to run a war, we should have run it right," said John Rector. "Use everything you have and get it over with." Some here, like Clark Reitnauer, never lost their belief that maintaining civilization demanded our standing tough. Others, including Sam Beerman, and Rector at times, conceded that those who lost sons might have lost them in vain. But nearly all agreed we should "fight it all the way or pull out."

The sons of Hanford either received student deferments or went in with what at first was gung-ho enthusiasm but became, as the war dragged on, cynicism. The earnest patriots hoped to re-create their fathers' triumphs at Saipan, Anzio, Iwo Jima and, in a sense, at those first reactors by the Columbia. The rowdies expected an expense-paid party where they could deal dope, shoot guns into the jungle and screw the Saigon women. The new longhairs thought dying in the jungle was stupid, but although a few fled to Canada, fought successfully for CO status, or even went to jail, most lacked the support that might have allowed them to stand and challenge the military.

This is not to say Richlanders unanimously accepted military interventionism and related domestic authoritarianism. In the postwar period Richland developed an active United Nations association and an American Civil Liberties Union chapter. Its local Democratic organization supported the treaty banning aboveground nuclear tests. When Adlai Stevenson came to Kennewick in 1956, his audience of 1800 (the largest of any west of the Mississippi) was comprised to a large degree of Hanford scientists who respected his intellectual abilities.

During the generalized Vietnam-era revolt there were a few manifestations even here. Tri-Cities blacks paraded to the Pasco City Hall and, together with white supporters from the local CORE and NAACP chapters, sang, "We won't let Ed Hendler [the town's mayor] turn us around." They challenged Kennewick's housing segregation, and the unwritten law that they had to be out of that town by sunset, and protested when Alabama Sheriff Jim Clark came through to tell "the real story" of the problems down south. As early as 1965 a chemist named Archie Wilson wrote the first antiwar letter to the *Herald*—and though few initially shared his stand, Tri-Cities Democratic caucuses soon began criticizing America's involvement. Although the activist group was small, they worked on the McCarthy and Kennedy campaigns, joined with their Oregon neighbors to keep nerve gas from being transported to an army depot in the nearby town of Hermiston, and even held a twelve-person overnight vigil in front of Richland's Federal Building.

But most Hanford people remained unaware that, aside from a bit of draft board counseling, local dissent even existed. At work or in the bridge and Kiwanis clubs they performed their normal activities and discussed politics little more than they ever had. Even when they tried to grapple with change—like the members of a small Mormon group who met almost clandestinely to talk about social issues—the conversations kept steering back to such safe topics as sports, vacations or the quality of the cookies and coffee at the meetngs. In the words of a young activist who eventually became a state representative from Seattle, "People let you have your say, then got right back to business."

Although Tri-Cities radio stations, to take one example, remained so frozen in time that they stayed with Tommy Roe, Neil Diamond and occasional soft Beatles cuts while others around the country were playing Bob Dylan, Aretha Franklin and The Stones, cultural unrest reached the young far more than organized politics. Some of the resulting rebellion was flip: A group of smartasses chanted "Hiroshima, Nagasaki, Hiroshima Rah! Col High Bombers Hah Hah Hah!" at a football game and were nearly beaten up for their creative efforts. Some was more earnest: an unsuccessful attempt to change the Columbia High team names to something less militaristic. But for the most part the kids turned their alienation inward, with the drugs that Barbara Giroux from the bridge club referred to in admitting: "We had a bad situation here in Richland. We were a sweet little community, but we didn't escape it."

Doreen, who was in on the conversation, responded by describing her night of terror when she ran into a group of teenagers gathered by the Columbia in Richland's Howard Amons Park. "They were hippies, street people. None took baths except by walking in the river with their clothes on. They were not local people who did this. I could tell by the way they looked that they were from Seattle."

But the kids were from Richland, Kennewick and Pasco—as were those whose 1970 drug busts in Pasco's Volunteer Park set off two nights of rock throwing, window trashing and police use of clubs and tear gas. Rick himself graduated from high school hallucinating on psychedelics, along with a third of his 560-member class—and within three years he knew almost twenty people dead from overdoses and suicides. Virginia attended a formal dinner once with a couple who'd found out the hour before that their son had dropped out of college four months before, and they smiled politely while raising empty forks to their mouths. The son of another friend flipped out after too many acid megadoses and ended up for years in the local mental institution.

Try as they could to understand attitudes disdaining all they had worked for, the old hands could only shake their heads, as Virginia's husband Lester did at every longhair, and exclaim, "They need a good Depression to straighten them out." Or view the period, as one physicist's wife did, as a time "when nobody wanted money or practical jobs; they all wanted to be social workers instead and live on nothing and change the system." The generation that founded Hanford, as Virginia said, "took our kids through PTA, Boy Scouts and Girl Scouts. We supervised their school reports and got out old magazines to help them when they needed pictures for assignments. We followed all the right procedures from all the best child-rearing manuals. We couldn't handle their ending up like the kids on TV."

Although unrest and social breakdown shook the Hanford community, it was too distanced to leave the old hands with anything beyond a sense that a distressing period had briefly intruded upon their lives. The community of dissent was too small, immersion in ongoing Area and domestic labors too strong. As Hanford's children could and often did end up leaving, the community's basic assumptions were challenged considerably less here than elsewhere. The kids who stayed eventually claimed the private world compensations of drugs and sex. The most fundamentally critical aspects of 1960s opposition culture dead-ended here.

Because Rick Price drew a high lottery number, he didn't have to worry about the draft. I met him in Seattle, where he was studying history at the University of Washington, and moving furniture for Bekins to pay his way through. In some ways he enjoyed being cynical regarding Hanford. He'd tell me about the brief period he spent working construction at one of the WPPSS reactors, and how his co-workers got stoned in the containment vessel and went on Nevada coyote hunts where they gunned down the animals from the air with semiautomatic weapons. He mentioned friends whose garages were packed with expropriated wrenches, drills, saws and even full-scale welding outfits. He remembered stealing the Department of Energy flag when he became fed up with the new hustling attitudes and finally quit.

When Rick arrived at the university in the early 70s, some of his classmates reacted strangely when he told them where he was from. Rather than just another unexceptional place of origin, they considered Hanford a demonic fountainhead, a symbol of a broader war culture—and his father and the other old hands were bomb makers.

For all Rick's cynicism, the image of his father as a Strangelovian warmonger was ridiculous. John was an ordinary citizen who did his job, tinkered in his workshop and worried far more about the decline of pride in Hanford's younger workers than about the Great Red Threat. He let the generals and government leaders debate first-strike capabilities and theories of global counterforce. He wanted merely—as he would have under any system—to be given the tools to do his work and then be left alone.

Rick respected this, and was overjoyed when John announced that he had heard about a possible method to neutralize radioactive waste and make it entirely safe. Although the discovery wouldn't redeem the nuclear weapons Rick considered part of an absurd world power game, it meant the effort and inventiveness his father and his father's co-workers put into developing the peaceful atom might yet yield beneficial fruit.

Rick's older brother Mike considered working at the Area a higher-paid version of working at the local french fry plants. Like Rick, he grew up a normal kid, playing just as he would have anywhere else, and not regarding the Hanford environment as in any way special. He went through the Vietnam period receiving first a student deferment, then drawing a lottery number so marginal he spent a year worrying that he was going to end up in the middle of the jungle with an M-16 in his hand. He graudated from the local community college with the apparent options of following his father's path or leaving Tri-Cities, as Rick would the following year.

Instead Mike married a Pasco woman, the daughter of an appliance store manager and, working odd jobs around Pasco, had two kids, bought a home west of the town in the middle of some old grape fields, and tried to identify with a culture preceding the one that brought his father here. Mike's passion was rebuilding the silver Camaros that rested like supersonic fighters around his ramshackle house. If he had to work until the cars would support him, it would be as a grunt, simply drawing wages and logging time, not joining a technical team that required allegiance and support. Instead of waiting, bargaining, or possibly even bribing his way into a high-paying craft like the pipe fitter's union, he went down to the local Laborer's Hall, and took non-Area jobs until all he was offered one day was work doing minor repairs at the research, testing and fuel fabrication facilities of Hanford's "300 Area." If you turned down jobs too often you didn't get called again, so Mike went, figuring that at least it was construction rather than operations.

When Mike went in to work, it was the first time he'd been inside the reservation since a childhood tour. Because half his time was spent just standing around, he found dozens of unused rooms and storage areas where he curled up for hours reading while his shiftmates talked elk, four-wheel drives and football. He avoided hot zone work by delaying getting the special glasses that would fit beneath the contamination suit's breather mask. He wandered from building to building, stretching minor errands into half-hour trips, and, except for the top security areas, he went wherever he wished as an anonymous workman.

Mike worked there a year and a half hauling pipes, pouring concrete, building scaffoldings and carrying lumber. For a while it was tiring, but with all the exploring, almost fun. Then something started bothering him: When he asked people what they were doing, they could explain their particular systems or projects in all technical aspects but, because the nuclear systems were so huge and complex, they usually passed all his more general questions on to colleagues in other departments. One day he was in a room adjacent to a bank of hot cells where workers peered through leaded glass windows and performed remote control operations. Noticing a stack of *Playboy* and *Penthouse* magazines on someone's empty desk, he began leafing through them distractedly, and came across an antinuclear *Penthouse* article entitled "Our National Death Wish."

"It wasn't any mystic revelation," Mike explained to me later, while showing me various Hanford memorabilia he'd collected. "But I read the article, and it was a good one. Then I looked at the hot cells with all the instruments, the remote manipulators, and those nameless substances inside. I decided this was one hell of an environment in which to read about nuclear dangers."

Mike began thinking about how easily he'd come to view working here as nothing special. When a brief layoff hit a few days later, he resolved not to return but to build up his Camaro business instead. Admittedly sales depended to a large degree on the health of the atomic economy, and admittedly Mike was tinkering with the cars not so differently than his father had with his atomic projects. But he wasn't feeding directly from the government trough, and he could at least claim the Camaro effort as his own.

At a rundown house between the southern edge of Kennewick and the nearby town of Finley, I talked with three more of Atomic City's children. Except for the Hanford badge hanging from twenty-year-old Cindy's silver neck chain, they could have been any scraggly, wholesome, freckled country kids anywhere. While Cindy went inside to get beer, I sat with her twenty-two-year-old boyfriend Brian and their housemate Larry on the faded salmon planking of their porch. Larry, a twenty-four-year-old carpenter at the Area, remembered how his radiation monitor father was periodically called in the middle of the night because some employee had tracked home hot materials on his shoes and the monitors had to go to their house and rip up the contaminated part of their carpet. "But it was nothing special," Larry said, "just a job he did like somebody else might work in an auto shop."

Brian described how he placed uranium pellets into zirconium assemblies at Exxon's fuel element fabrication plant, just south of the reservation boundary, and how he had to shower every time he touched his hands to his face. Cindy, who'd returned with the beer, said he should shower more often anyway. She then announced that this afternoon NRC inspectors had visited the WPPSS subcontractor she worked for "and told us we were all doing a really good job." Though she stated this proudly, as if it made her part of a top-grade team, Brian laughed and responded, "Yeah, you're giving a real 100 percent. I bet they just liked your pretty face."

"You chauvinist, Brian," said Cindy in a girlish voice. "You know I always work hard at my job."

"I guess maybe you do," said Rick, "but I remember watching the Three Mile Island news with my fingers crossed, knowing that those operators were just piling fuckup on top of fuckup."

Brian hitched up the coveralls which made him resemble a husky, broad-featured Huck Finn. Sipping his Bud, he said Hanford and projects like it were just favors passed between politicians, and that the reactor operators might train forever but there'd still be accidents: "I go up and down for nukes and against them. But I remember this pipe at one of the WPPSS plants that looked perfect on the prints but went straight into the ground going nowhere. Sometimes people even weld bolts in instead of screwing on the nuts, just because it's easier and faster. When I think of that stuff, I start sliding down the pole to becoming anti-nuke."

"The pipe must have been the secretary's fault," said Cindy sarcastically. "She probably typed the plans wrong like they always do."

"Yeah, that's usually it," agreed Brian, except that he was serious and she wasn't; so she told him about a woman engineer "who outranks nearly all the guys, so don't you give us girls any shit."

"Bet she can't even pull slump—or know what it is," said Brian.

"Hell, even I can do that," said Cindy. And she pantomimed working her hands one over the other down a pipe's butt end: "Then you just go bam bam bam and that's it."

"Well," conceded Brian, "that's pretty good. How many times have I said I could teach you everything I know in two minutes?"

Cindy seemed content with her victory. Brian complained

that just because this place was "a pit" didn't mean they should use it to dump the Three Mile Island wastes which were now coming in by truck. Cindy said that whenever she asked critical questions, "the people on top always give the same raps," and that they never seemed to know what they were talking about. "But I guess," she said, "the world's supposed to end in hellfire anyway. At least that's what the Bible says to believe."

"Do you think that's true?" I asked.

"Well, I'm not sure."

Asked if he felt strange about building the plants, Brian answered: "I would at Satsop [an at times contested WPPSS site 200 miles to the west on Washington's Olympic Peninsula] or anywhere else they didn't want them. But this is a nuclear town and it always will be."

Cindy brought out another six-pack. Brian took his shoes off, leaned against the porch railing, and looked off, imagining a distant confrontation: "If I was back East and they were building reactors," he said, "I'd be throwing rocks, climbing fences and getting arrested. Here, where else are you going to work? I tell you, I don't like the FFTF, though, and when they start that sucker up I'm moving.

"Yeah, we'll leave when that FFTF starts up," Brian reiterated. "When they call it Fast Flux Test Facility, remember that's *Test, Test, Test*. If it blows up we'll be part of the test that failed."

Cindy repeated her contention that "we're all going anyhow." I walked out past her VW van with its Greenpeace SAVE THE WHALES sticker, and drove off beneath a pale orange moon.

5

Boomtown Cowboys I

> "I fucked the company. The company fucked WPPSS. WPPSS fucked the government. And the government fucked me."—graffiti at one of the WPPSS reactors

As a government industry, Hanford remained vulnerable to changes in policy. In January 1964, with N reactor's generating plant not even completed, the Johnson administration—deciding it had sufficient plutonium stockpiled and facing general pressure for disarmament—curtailed Hanford production by 25 percent. Three reactors were shut down in the next two years. In the period between 1966 and 1971 all the rest except N ceased operation. N itself ended up primarily producing not weapons-grade plutonium, but the less pure fuel and research grade which allowed fuel elements to be changed less frequently and more electricity to be generated.

People began worrying about the cutbacks around 1969, by which time five reactors plus a number of the processing plants had closed, the shutdown of another three was imminent, and the layoffs had already begun that would drop operating employment from 8,500 persons in 1967 to 5,500 just four years later. Although some of these cutbacks were accomplished through general attrition, the community was young enough that few were ready yet to retire, and many had no choice but to leave town—migrating to the Southern California facilities of General Atomics or to the East Coast naval yards that built the nuclear-powered submarines. Others stayed waiting and hoping, as one engineer said, "that because they'd been here twenty years the government would find some way to take care of them." Those who kept their jobs or transferred from the closed down reactors to still operating projects assumed the layoffs wouldn't affect them, or else—like Virginia's husband Lester—insisting they always knew that whatever the government gave it could also take back, proceeded to send out resumés to other nuclear enterprises around the country. Hanford workers even received paid time off for interviews. But most here did everything they could to remain in the town the atom built.

The initial cuts did evoke action by Tri-Cities civic leaders, but what catalyzed the community was Nixon's proposal in 1971 that—again due to the plutonium surplus—N reactor should be shut down as well. The loss of N meant a possible end to the bulk of operations and threatened to make Richland, which still had essentially no industry outside the Area, a town nearly as dead as White Bluffs and old Hanford.

To respond to this threat, the community called in all the clout it had with Washington Senators Warren Magnuson and "Scoop" Jackson, as well as with former project head Bill Johnson, now serving on the Atomic Energy Commission in D.C. School kids—including Virginia's third-graders—wrote letters to Nixon explaining why N reactor was important and how terrible it would be should their families have to move. Adults wrote and telegrammed as well, lobbying in every manner possible until, for whatever reasons, the decision was reversed and N reactor stayed.

If a single individual could be considered responsible for ending Hanford's brief economic crisis and for creating the current boom economy—in which between 70 and 80 percent of Tri-Cities workers (and 95 percent of those living in Richland) either hold Area jobs or sell Pontiacs to, handle the banking deposits for, or teach the children of those who do—it was a short, goateed civic mover named Sam Volpentest. This eastern Washington power-broker came to Richland in 1949 from Seattle, started the town's second tavern, Tri-Cities' first locally headquartered bank, and—to promote the region—formed the Tri-City Nuclear Industrial Council. TCNIC's first task came when General Electric decided to pull out from their AEC contract. There was a chance the entire Hanford complex would fold at that point as well, but Volpentest used contacts with Jackson and Magnuson to line up a new consortium that came in between January 1965 and July 1967 and included Douglas United Nuclear (a joint venture of Douglas Aircraft and United Nuclear Corporation) at the reactors, Isochem (a project of Martin Marietta and Uniroyal) to do the chemical processing and waste management, Battelle Memorial Institute to handle research, and ITT Federal Support Services for a variety of miscellaneous tasks.

When I met the now seventy-eight-year-old Volpentest in his Richland office, he recounted how, after Nixon decided to shut down the old plutonium reactors "and we thought we'd had it," he and *Tri-City Herald* editor Glenn Lee (who joined him on TCNIC, along with the paper's publisher) decided to go after a government linear accelerator. "The officials kept telling us 'forget it,' " Volpentest said with a veteran's pride, "but we made eight Washington D.C. lobbying trips, and when Congressman Mel Price swung the contract for his Illinois district, they asked if we wanted the breeder as consolation. Lee said 'Don't do us any favors,' but I insisted even crumbs were something, and suddenly it was being built here instead of Idaho Falls. Then WPPSS had problems at a planned reactor site on the coast when the environmentalists got all hot about Indians, fish and wildlife, so we convinced them to give their plants to us as well."

The phone rang and Volpentest talked high-priced real estate for a moment: "Do I have pull with the city engineer? Oh, I know him, but I'm knee-deep in alligators this week." He searched the file cabinet beneath a painting of a massive dam spillway and a photo of himself with Jackson, wearing white hard hats in front of the newly built FFTF. He gestured quickly with a carnival barker's constantly moving hands, and turned to me smiling, explaining apologetically, "They all want favors."

When I asked about the then recent Three Mile Island accident, Volpentest said, "You bet it was a major incident. But that's the fault of the utility, who started up before everything checked out. If you don't run things right it's exactly what the anti-nuke people need." I asked about an NRC investigator's statement that the reactor was out of control and that only luck prevented a meltdown.

"That's asinine," Volpentest said, momentarily angry. "If he's on the NRC it's like the blind leading the blind because if anything the accident showed the backup systems worked." Volpentest slowed, cooled down, and smiled again. "But I don't want to get into an argument about Three Mile Island. I'm not a technician. I just take what I read in the trade magazines and the word of people I trust. I believe the designers who say our reactors are safe—if I didn't I wouldn't let my family live here—and I have faith in the wisdom of congressional budget-makers who've given us funds to continue development."

The phone rang again and Volpentest explained to someone, "Tell the buffet manager to give FFTF three tables near Magnuson."

He smiled almost sheepishly at his latest atomic horse-trading, and repeated, "My wife thinks they just call because they want favors, but I guess by now it's in my blood." He joked about how, after he whipped a jaw cancer that still pockmarked his face, people thought he survived by being irradiated, "but that was nonsense—I just had a good doctor, I was wired together for months. Now I just want to keep on going and living." He said Tri-Cities was founded on a single industry but was trying now "to be a real American city." I asked if he thought this would really happen.

"WPPSS and the breeder really turned things around," he concluded. "We're striving now to become a total energy center where—though I may not live to see this completed—we'll have fifteen to twenty more plants, 150,000 to 200,000 people, and enough processing and storage facilities so raw uranium can come in, electricity can flow out, and that will be it. It seems Oregon doesn't want more reactors. And maybe California with that crazy Jerry Brown doesn't either. But we have the skills and the enthusiasm. We'll be nice and sell them what we produce."

When I visited the WPPSS sites, their cooling towers, instead of being the familiar Three Mile Island hourglasses, resembled four-story stacks of giant black poker chips topped with suction cups. WNP 2 (the official designation, standing for Washington Nuclear Plant) could have been a futuristic Sears building. WNP 4, just beginning (WNP plants 1, 2 and 4 are located here, numbers 3 and 5 200 miles to the west, by the Olympic Peninsula community of Satsop), displayed only its skeleton of steel reinforcement bars, which formed a slate crisscross blanket over the rust-colored metal of the inner containment wall.

WNP 1 is a 1250 megawatt Babcock and Wilcox reactor which WPPSS hopes will go on line by July 1985. On a lot the size of a football field outside the plant, eight-foot pipes curved and twisted like giant multicolored earthworms. Adjacent to this lot, laborers in flannel shirts, down vests and red hard hats carried planks and buckets of nails past shacks and portable trailers housing lunchrooms, toilets and the on-site offices of support personnel. Cranes sat ready to lift girders or equipment sections. Except for the reactor dome's final section, which rested on the ground as if it were the orange skullcap of a giant, this could have been any megatech project anywhere.

The WNP 1 containment building was a standard, though as yet incomplete, egg-shaped dome, sitting to the right of a rectangular four-story building that housed the plant's turbines and generators. On the containment wall's outermost layer, fist-thick rods of the ribbed reinforcement steel called "rebar" curved up like 20-foot tusks from the concrete in which they were anchored below. Green-helmeted structural ironworkers wired in more rebar horizontally, forming a grid over which the next level of concrete would be poured. On a ramp, leading up to the 235-foot-high building's entrance, a yellow crane called the FMC Link-Belt lowered a series of thirteen-ton tanks (for demineralized water) into the frame built to hold them. Two men stood on top of planks to guide them in by hand. One, wearing a pipe fitter's blue hard hat, served as flagman by directing the crane operator with hand signals. The operator sat in his cab between each pick (the shorthand term for lift) and sipped coffee.

The reactor pressure vessel, itself 43 feet high and weighing 1100 tons, had been lifted in a single pick over the building's open top by a specially built crane called the Transilift, which was the largest mobile crane in the world. With its load counterbalanced by a rear boom almost as long as the 400-foot front one, and by stacks of concrete blocks, each weighing 36 tons, that rested on side extensions resembling stubby wings, the Transilift overshadowed the reactor like the neck of some frozen blue reptile.

My PR guide and I entered the containment building past spotlights used for night shift work. We watched another smaller crane—itself carried by the Transilift to its perch on the building's rim—swing a long arm across the diameter of the space and move a bird cage containing six men from the ground to a scaffolding halfway up the wall. We descended on catwalks and iron stairs along the space between the outer and inner containment walls, past pipes and conduits which wove through various portals like cables hanging from the hull of a ship. On one wall, above a beam on which lay old copies of *Guns and Ammo* and *Outdoor Life*, someone had written the title of Ian Dury's punk anthem: "Sex & Drugs & Rock & Roll"; on another, farther down, the admonition, "WPPSS needs Lerts. Be a Lert."

At the bottom now, I looked up as the vessel rose above me, a huge black obelisk thrust up from the ground. Beams crossed, forty feet up, to brace it. A steam pipe larger around than a railroad tank car, covered with a plastic tarp, ascended to my left. Ports at different levels allowed other pipes to carry heated steam to the adjacent turbine building. Except for a hum from scattered welding machines, the complex was strangely quiet, and a thin November drizzle left mud and grit on the cold concrete floor.

As I passed three pipe fitters welding, I remembered a Philip K. Dick novel in which the plumbers' union takes advantage of the need for constant canal upkeep to run the planet Mars. WNP 1 contained over one hundred miles of different pipes, many of which will hold radioactive liquids at temperatures of up to 580 degrees and pressures of a thousand pounds per inch. With faulty connections risking, despite backup systems, potentially major accidents, it's no wonder Local 598 of the Plumbers and Steam Fitters (called simply the pipe fitters) is considered to own Tri-Cities. A bathroom graffito explains, "Welders are like whores, always hollering for more heat, more rod and more money." It's a Hanford commonplace that food prices go up each time the local renegotiates a contract. Their $19.40 an hour 1979 scale (increased to $24.18 by the time I returned a year later) was the highest of any craft here, and one of the highest of any plumber's and steam fitters' local in the nation.

The fitters in front of me now, one grizzled old hand and two young guys, stood on a scaffolding platform several levels up in the reactor. Lifting a 300-pound pipe section into place with pulleys and chains, they lowered the section onto a swivel stand that held its weight. One of the young men, who had a long blond ponytail, shone a portable light as the older hand measured carefully, scored around the surface with a chisel, then explained, when I asked what he was doing, "Got to miter it up." The other young worker picked up a welding torch, knocked it against the scaffolding rail to clear out any dirt, then produced a burst of yellow flame, which he honed to an even hotter blue by adjusting the gas mixture. Pulling down his goggles, he torch-cut following the scoring marks so the pipe section would fit snugly with the adjacent one.

Outside, the Link Belt crane was helping disassemble a bridge that led from a concrete ramp to the containment building entrance. As the crane slowly removed one of the bridge's four-ton girders, a half dozen ironworkers nursed it around, a compliant but unpredictable airborne steed. Two men in yellow electricians' hats walked by with packages of light bulbs. A young laborer, sporting shoulder-length hair and a knit sweater from which Bugs Bunny looked out with a stoned anarchic grin, carried boards down a stairway on the containment structure's left. Another laborer retouched the red paint on a crane hook.

When I talked with a young Quality Control inspector who, wearing the white hard hat of a technician or supervisor, checked the green QC tags on each rod in a bundle of rebar, he described his work—"enforcing each petty requirement"—as "the asshole job of this whole operation." A brown-hatted carpenter recalled a recent "wobble," or task jurisdiction dispute. His union had installed a set of imbeds (pieces of conduit which you placed inside forms and then ran electrical wiring through) and the electricians ripped them all out, claiming it was their job. As I left, a crane swung a one-ton concrete block in a graceful, almost weightless arc, behind the building edge and out of sight.

When ironworkers complete a major project—whether a bridge, a high-rise building, a dam, a reactor containment building or a missile silo—they often steal a pair of panties from one of their wives or girlfriends and hang them on a final girder christened "the last piece." But it seems at times no panties will ever get hung from the WPPSS sites.

WNP 1 was budgeted in 1973 to cost $660 million and was expected to be completed by 1980. The reactor's direct construction costs are now projected to run $3.8 billion (or three times as much, counting interest on issued bonds) and its announced completion date is 1986. Along with the other two Hanford WPPSS reactors and the two at Satsop, the total effort's estimated costs have increased from $4.1 billion to $24 billion, and now surpass those of the Alaska Pipeline.

The roots of these projects began when the Bonneville Power Administration (BPA) was formed in the 1930s to market the electricity from a series of dams, including Grand Coulee, which the government was building along the Columbia. BPA sold its power first to public utilities, next to directly contracted industries such as the major aluminum producers who moved to the region at BPA's invitation, then to private utilities, and finally, in times of surplus, to utilities in less hydro-wealthy states (California was one). In the late 1950s Washington utilities decided that the growing need for power would eventually outstrip the facilities of existing dams. In 1957 they formed WPPSS, as a consortium of Public Utility Districts and municipal power systems, for the purpose of producing their own power and distributing it over the existing BPA transmission grid. They built one small dam at Packwood Lake in the Southern Cascades, then initiated N reactor's Hanford Generating Project (HGP) in 1963.

For years HGP was, along with the Packwood Dam, the sole WPPSS enterprise, but in 1969 the Joint Power Planning Council—a Northwest group including 104 PUD's and municipal co-ops, four private utilities and the BPA—asked WPPSS to sponsor first one, then three nuclear plants to provide the region with additional generating capacity. WPPSS responded by enlisting its nineteen-member utilities, four participating municipal systems, and ninety-two additional public and private utilities in Washington, Oregon, California, Idaho, Montana, Nevada and Wyoming to invest in what eventually became five plants. They drew up a law called a Net Billing Agreement which Scoop Jackson lobbied through Congress in 1970, mandating that participating utilities pay for the cost of all electricity produced by the initial three reactors (plants 4 and 5 were separately financed through direct investment by some, though not all, of the member utilities); Northwest ratepayers would be required to pay all costs of building, even if the reactors never yielded a kilowatt.

The Net Billing Agreement has allowed bonds for the first three WPPSS plants to carry top-level AAA ratings and therefore sell relatively easily; over half the municipal bonds offered in the United States are those of WPPSS, and the size of the debt for the five reactors is exceeded only by that of the federal government. But a family of four in a city or utility district committed to all five plants is responsible for a median indebtedness, just for projected construction costs, of between $27,000 and $30,000—to be paid off through electricity rate increases scheduled to jump fivefold in the coming decade. BPA, after raising its prices only three times in forty-two years, all in minor 10 to 14 percent rate hikes, has announced successive 88 percent and 53 percent increases in a recent six-month period. The time when the Northwest's hydropower gave it the cheapest energy in America is nearly over.

For craftsmen used to working themselves out of jobs as they complete each project this "construction gig that never ends" is a seemingly infinite source of wealth. When I met Lou Hansen, a twenty-eight-year-old pipe fitter who works Quality Assurance at one of the WPPSS sites, he was quick to test me by grinning sardonically and asking whether I was like other journalists who viewed Richland as a town ringed by ominous cooling towers and who filed reports of catastrophic accidents "every time someone trips in a Hanford parking lot."

Lou's job was fairly easy: checking welds and varying installations made by the other fitters to ensure they were done properly, located correctly and followed all specifications from the plans. Since it wasn't production work, the demands were intermittent; on some days he'd spend all but ten minutes racing around and on others barely ten out of the chair where he sat and read when things were slow. The year before he had drawn double time (since changed to time and a half) for Saturday work during which he consistently had nothing to do, but was required to be there in case the welds passed through any of the up to a dozen different stages that had to be checked.

Despite his own comfortable situation, Lou spoke of WPPSS as a "hustle-world" where fights and strikes erupted constantly over which union would send their men to carry some girders or erect some four-foot scaffolding section; where foremen stole arc-welders and drill sets by hiding them beneath tarpaulins in their pickups; where older craftsmen took their kids straight out of high school, gave them some training and got them welding jobs making $750 a week—from which there was "no place they can go but down."

WPPSS work was hard and, as with that on any major project, risky; falls and other accidents had already taken seven lives. Because the summer desert baked the containment vessels up to 130 degrees inside, it was largely a young man's game. With much of the work both difficult and demanding of skill, Lou resented those "who call the technical people professionals and imply that we're some kind of amateurs."

But as there were no on-site radioactive materials, Lou explained that the only difference between constructing these reactors "and putting together some auto body or dry-cleaning boiler" was an invisible danger which most of the workers wouldn't even stick around to see. Also, since the money was so good, they tried to prolong the flow as long as possible. It was an understood maxim that "nobody can lay pipe too slowly."

Lou lived on Kennewick's Clearwater Avenue, a street whose transition from farm road to drag strip echoed the town's WPPSS-fueled growth from 18,000 inhabitants in 1975 to 33,000 just four years later. Clearwater began with rows of McDonalds, Dairy Queens and overpriced Edwardian-style steakhouses; it continued past tacky six- and eight-family apartments and past cedar arcades, such as the Clearwater Industrial Center, that housed the realtors, upholsterers and insurance brokers who had come to serve the waves of new residents; it ended amid open sagebrush with the 40-acre Columbia Center shopping mall.

Lou's own apartment was in a new complex called the Marina Vista after an algae-covered "lagoon" between the parking lot and the surrounding desert. The building alternated facades of imitation stone and imitation brick with greenish-yellow paint. With an entry through an arbor vaguely resembling a tiki shrine, it was a Club Med version of all the town's new instant oatmeal complexes.

Inside, the apartment overflowed with an array of expensive cameras, stereo components and other electronic gimcracks capped by two mint-condition antique pinball machines. Looking like a sharp lean mechanic in jeans, a blue madras shirt and Frye boots, Lou sat on a new maroon and orange love seat next to a stack of magazines that included *Playboy, Omni, Mother Earth News* and his girlfriend's copies of *Cosmo*. He said that if I thought this apartment complex was strange, I should have seen the one before which had green and blue doors, globe lamps, a motel-style layout, "and felt so much like a Travelodge I woke up each morning wondering what time I'd have to check out." He talked of co-workers who sent all they earned to families back home, and ones who spent it as fast as it came in on Harleys, boats, "any sort of cars as long as they are new and nice," dinner seven nights a week at the local steak and prime rib houses and the full bore weekend trips to the coast.

"From what I make," Lou said, "which was $32,000 last year, I give lots to the government, save a bunch, and buy a few antiques like these pinball machines. I'm no big consumer, though; I just bought an old Ford Catalina for what some people would spend each month paying off their new four-wheel drives. And I like having some cushion in case things slow down."

Lou's apartment was well kept up, but when I visited his co-worker Dave, I entered past kids playing listlessly around a dirt sandbox and broken swings, past a torn screen door and past a bathroom where the "decorator-patterned" formica was cracked and crumbling. The complex, called the Park Lane, took up a full block in a backwater area of Richland. Its two-story buildings wobbled on stilts above concrete carports. Some scraggly grass constituted the only Richland lawn left uncut.

The Park Lane, Dave explained, was—though most of its tenants were white—as close as Richland got to a ghetto. It was a flophouse for mobile boomers on their way in or out of town, and the residential equivalent of the rows of ministorage buildings that sprang up everywhere here to serve a culture constantly on the move. Dave had lived here once before, with his wife Cindy and her fourteen-year-old son (they were both in their early thirties), when he was bouncing—"just like a hired gun"—between work here and jobs in Texas, Pennsylvania, Florida, Montana and Mexico. While Dave admitted the place was "somewhat ugly," in a rental market of 1 to 1½ percent vacancies at least it often had empty units.

Dave was big, 6′ 4″, 240 pounds, and wore his hair in a long dirty ponytail that would have passed muster in the seamiest acid zones of New York's Lower East Side. When a South Carolina ironworker dropped by with his wife, the men joked about ripping off the "Stalag 13" tavern where Cindy was employed and where WPPSS workers cashed their paychecks ("Oh no, you don't," she said. "Not so long as I'm around."), then switched to tales—high dollar-powered variants of the sort recounted by hard-ass braggers anywhere—of their Porsches and Harleys, of star hydroplane racers, drunken fishing trips and shiftmates who avoided doing all work while drawing full pay. The women planned the night's chicken barbecue and commiserated with one another about the difficulties—because their families were back on the other coast—of finding reliable baby-sitters. The two sexes connected briefly when Dave recalled finding old Gene Krupa and Benny Goodman records in an attic, when the South Carolinian recounted exploits of small-time drug smuggling and when they passed joints of $150-an-ounce grass that Dave explained to me was "not terrible, but certainly not the best." In the background a color TV—sound turned off—flashed pictures from an African safari show and commercials of a man waltzing ecstatically with his new electric razor. On the quadraphonic stereo, the Jefferson Airplane sang "One generation got old. One generation got soul . . . Need a revolution, got a revolution," from Dave's battered copy of the "Woodstock" album. Undone laundry lay everywhere. When his son wandered in smoking Marlboros and wearing a sleek cream-colored body shirt, Dave told him, "You look like a real white-hat."

When I asked Dave what he felt about nuclear power, he answered, "Well, when a man came through who'd been at Three Mile Island I didn't shake his hand," and explained that he wanted "out of here first chance I get."

Steve, a rangy Montana friend of his who'd dropped by, said dryly, "I've seen you pass up opportunities."

"Well, they'd have to be good ones," answered Dave, laughing. "Long hours with high bucks, 'cause you know I'm just in it for the money."

"You know sometimes Dave causes shit by doing his job too well," said Steve, now defending his old friend. "He may smoke dope with the best of them, but if a weld doesn't cut it [Dave, like Lou, worked QA] he'll have them do it, redo it and redo it again. And even then he still might not let it pass."

Dave went to the back bedroom to exchange his flannel shirt for a chopped-off jean vest. Cindy told me, "It's hard now that he's making so much more here than he ever did before," and described their hope "to save up a bunch of money and live on a Mexican beach for a year." Dave returned, agreed nothing at Hanford had reassured him about the nuclear industry and said he'd still support any town that didn't want reactors in their backyard. "Maybe," he said, "I'd be happier going back to fighting forest fires for six bucks an hour. But while I marched a lot in the sixties and still believe all that idealistic Mahatma Gandhi stuff, it kind of dries up when you land on your feet, have two kids and realize you need money for gas, weed, rent and wine. I figure the reactors will get built—or maybe endlessly delayed—whether or not I work on them and get my fair share of the bucks going round. And at least I'll check the welds more carefully than someone who doesn't give a shit."

The next day Lou and I drove to an old Irish bar in Pasco—where he'd grown up across the Columbia from Kennewick—and listened as a truck driver described nuclear power as "just another big-money scheme" of the giant oil companies and of the line of corrupt politicians who succeeded the "honest and plain-talking" Harry Truman. "But it's here," the man said resignedly, "so I guess we must need it."

Pasco's population of 16,000 is only 1,500 larger than it was twenty years ago. The town hasn't been as heavily influenced by Hanford as have Richland or Kennewick, but it hasn't escaped the nuclear era altogether. Farmers still park their ancient pickups and battered Chryslers outside houses hand-built many years ago. Tape still forms gray mosaics on the unrepaired windows of cut-rate garages. Women still count out pennies to buy packages of hairpins at the local drugstores. But some of the good ol' boys sitting on the stoops drinking beer are Hanford construction workers, and when they take snapshots of their wives, their girlfriends or their new four-wheel drives, their Nikon lenses and cameras cost together over $2,000.

Except for Tri-Cities' only porn theater (a former vaudeville house where Lou used to pay twenty-five cents to watch Disney matinees), Pasco's downtown is now badly eclipsed by Columbia Center. We drove east from the old stores that reminded me of empty Greyhound stations, and passed streets of white clapboard houses with sagging gables, gardens that old Chicano women (come with their families to work the nearby fields) had transformed into bursting celebrations full of cataloupes, jalapeño peppers and orange and yellow marigolds, and—as we crossed the railroad tracks to the black and Chicano ghetto on the far side—weathered shacks that could have been transplanted from rural Mississippi. On the town's outskirts, in an interracial border zone, we stopped at a rambling old A-frame house where three generations sat talking on the lawn. Mark, a gentle twenty-one-year-old who worked at the Mojonnier & Sons potato processing plant, told how his Indian grandmother used to pick him edible wild berries and cinnamon grass. He said he might want to go back to Hanford, where he had worked in a uranium preparation plant, then looked off silently as if the wealth the Area promised was that of a seductive Emerald City just over the horizon. "It would have to be an outside job, though, like a carpenter," he said emphatically while his three-year-old daughter offered him leaves from the scraggly bush they called "a chickenfeed tree." "I just can't take those nose-in-the-air Richland people, and though I'll work near nukes, I think they're going to be the end of the world. You know—like the way the book of Revelations says. I'm not religious or anything, but I just have a feeling that some time one of those chain reactions just isn't going to stop."

As we left, Lou said Mark's fear was "clearly due to all that radiation corroding his brain. You'll feel it too if you stay here a while: a sneaking suspicion that we're all really living inside a Gahan Wilson cartoon." Looking at the streets we were passing, he recounted cruising them as a teenager from Pasco to Kennewick and back. Kids from both towns would repeat this nine times per Saturday night, "just like doing laps in a pool," but no one would ever hang out in Richland.

When I asked why not, Lou called its people "snobby, like that farm kid said," and largely just "an Ozzie and Harriet culture," where Dad went off to invisible work and Mom stayed home to do the dishes and manicure the dog, where the worst social problem was acne, and where even the few blacks and ethnics were "rinsed almost white with scientific training."

We drove through Richland now, on a perimeter road that wound past the Bible Way Revival Center and the Sunset Park cemetery with atomic symbols on its gates. I remembered an old Pasco man's comment: "Richlanders are like seagulls.

They're protected by Uncle Sam, they get everything for free and they crap wherever they want." After continuing on to the WPPSS site, Lou pointed through a chain link fence to show the project where he worked. The sprawl of trailers and outhouses and "Dairy Queen shacks" was "for the armies of people who fill the water jugs, set up the picnic tables and run the trash compactors." The circular cooling towers, should the project ever be abandoned, could become great condominiums "since they have balconies already in place and set to go."

Lou stopped the car to deliver a small machine-unit to a friend who owned a twenty-person welding shop on the border between the Hanford reservation and the town of Richland. This man, about to go out of business, explained that, because of the Area's exaggerated wage scale, he lost any decent workers or paid prevailing rates and consequently—even if the contract was for a job literally a stone's throw away in one of the Hanford complexes—got underbid by some $4-an-hour Seattle or New Jersey contractor. "When you add in dealing with taxes and regulations," he concluded, "small-timers like me really don't stand a chance here. So I'm leasing my space to United Nuclear and going into mobile homes and real estate."

Although Richland still hadn't developed a small business class and this independent shop was, therefore, an exception here, Lou explained how Pasco and Kennewick were at least partially self-reliant. "While we may work at the Area, that doesn't mean we always defend it the way *they* do," he said, referring to the Richland professionals and managers. He remembered doing some spot repairs once in an N-plant hot zone when the reactor was shut down for the summer. The radiation monitors clocked him. He wore a mask, protective overshoes, three pairs of cotton coveralls and a dosimeter which beeped as each particle hit. They assured him after he came out that the exposure was no more than if he'd been wearing a watch with a radium dial. "But I felt like the Lone Ranger withstanding a hail of bullets," he said. "I can think of plenty more fun ways to fracture my chromosomes."

"I guess I do prefer straight construction," Lou continued, "where if you make it through the day without busting your head you know you're okay. I do worry that instead of a single active plant 30 miles away, there will soon be three new ones less than 8 miles into the reservation. But while I can't vouch for design, in terms of day-to-day construction these reactors are built well. I'm all for wind, solar and small hydro projects, but you know being an American takes lots and lots of energy."

Stalag 13, the tavern Dave and his friends joked about ripping off, was an after-work stopping place located on the road running from the WPPSS and other Area sites to the three towns. Construction workers just off shift played the half-dozen pool tables, the twin pinball machines and the jukebox. (Stalag also had a wide-screen TV, electronic hockey and Space Invaders games, as well as a color organ that synchronized with the music at night.) The men lined the tavern's west wall to cash their weekly paychecks at a special window.

Chili, a hard-ass in a beat-up army jacket, invited me to sit down at his table, and began by telling me: "First thing you've gotta know is that if the management would only let us, we could build the plants in two years flat." Then a young bearded kid from another table passed and Chili yelled, "Hey Denny, you got anything for the nose?"

"No," Denny answered, "but I've got some great $250-an-ounce Sensimilla."

"Shit," Chili said, "I can get $95-an-ounce stuff that can knock you out, so if I pay $250 I don't want none of this 'two toke' shit. I should be able to just look at it and get high, because otherwise paying that much is just plain stupid."

Now around age thirty-two, Chili went to Vietnam after growing up in nearby Walla Walla, and ended up here after moving through the construction migrant's standard pinball carom circuit from project to project in state after state. When I asked if people here smoked dope at work more than they did elsewhere, Chili grinned with the look of a man tough enough to get the job done through rain, sleet, snow or self-imposed brain defoliation: "I got just as loaded building Lower Monumental Dam," he said, referring to a Snake River project 35 miles northeast of Pasco, "but not so much I didn't know what I was doing. Here I'll enjoy it if I can, which means I'm careful around machinery that could nail my ass. But if I'm just on fire watch—standing with an extinguisher to make sure some welder doesn't set fire to a bunch of concrete that can't burn anyway—well, flames look lots better when you're high."

Chili left, saying he'd be around the next night to watch the mechanical bucking bull at Kennewick's Staggerhorse Bar. "Now be sure to remember my name," he told me, "so all the fuckers I was with in Nam will know this fucker's still smoking dope."

At a nearby table of young carpenters, one named Dick proceeded to caution me that all pipe fitters were (in a phrase lifted from Hunter S. Thompson) "treacherous bastards." Dick wasn't treacherous himself: still somewhat baby-faced at twenty-three, just a smart and smart-ass young guy whose family had moved to Kennewick from Ohio when he was twelve, and who now, whenever he leaves town, makes "damn sure to say I'm from Cincinnati, Seattle or anywhere but this dump."

Dick commemorated this statement by buying us all another round of beer, and a guy named Greg with a huge shock of blond hair returned from the phone to announce, "Drugs are here." Though what he'd lined up was psychedelic mushrooms, Greg toyed with Hanford's bad-ass myths by looking intently at me, rolling up his sleeve for an imaginary needle and explaining, "It's Mexican brown . . . You know: everything to excess. It's in our contract that they have to check your bumps before you're hired, and if they don't measure up you get a special training film on how to shoot and snort."

Dick mentioned reading that Brian Jones employed a special assistant "who went with him just to score whatever he needed." Someone else mentioned that Jim Morrison's birthday was on the tenth of the month. We talked about The Doors, about white blues player George Thorogood, and about a British band named Rockpile whose show, Dick said, "was just about the best live concert I've ever seen."

I told Dick I'd seen them tour with Elvis Costello. He raved about the Costello album, "This Year's Model," then said, "But you know who really did the rebellious rock the way it should be was the Sex Pistols."

I laughed. I said I knew what he meant. As I imagined the band singing "No future for you. No future for me," from a platform hung high above the containment vessel by the Transilift, Dick continued, "Yeah, they were a great band if you really listened." When I asked how he got introduced to the music, he answered, "I just bought all the new albums with wild-looking covers."

Mary, a silver-blond waitress, wobbled by, lost as usual in her own private Quaalude-land, and said, "It's good you guys didn't come over for Thanksgiving because I forgot to turn on the oven for the turkey and Frankie was smoking that wacky tobaccy, and he got mean and started punching everyone."

The gang left after voicing a final complaint about the government taking nearly half their income, and Dick invited me to meet him later at Kennewick's Cosmo Angus Bar.

I stayed talking with a kid in a Vette cap named Mark, who still had 50 miles to drive. He lived next to the Columbia, in a house he'd bought in an upriver recreation development just the other side of the area. The development was called the Desert Aire (on its entrance sign the "D" in "Desert" puffed out as if it were a balloon spinnaker), and Mark drove from there each day in a Honda Accord bought for routine runs because the Corvette, "my expensive toy," was too greedy on gas. He lived there because it was cheap, relaxing and low-keyed, plus—as he admitted with a laugh when I asked about safety—upwind from the reactors which he hoped would take so long to build that he'd be gone by the time they went on line.

While The Kinks sang from the jukebox about living a "Low Budget" life, Mark went to the bar to get a hot dog, then returned, sipped his final beer and recounted one of the scatological maxims that sprang up all around here like desert sagebrush. "An electrician I was working with told me life is like a shit sandwich," he said. "The more you have, the more shit you get. I've also heard it another way—that the more bread you have, the less shit you have to eat. The second one's probably more true, but if you follow it you might as well be a businessman or a lawyer. There've been lots of places where I've worked my ass off more than here, and sometimes it does start to get to me. But what I'd really like to do is see if the industry keeps booming and my land upriver increases in value. Then I'd sell it and move to Prescott, Arizona, where it's sunny all year with lots of water and fertile land. To live these days you need the green dollar. Electric companies keep boosting rates. Food costs more almost every month. But if I could get enough of a grubstake to buy something down there, maybe I could farm and have a solar thing and be almost self-supporting."

Megatech projects require megasized work forces: in the case of Hanford's three WPPSS plants, 9,000 construction employees. No town, certainly no small town, can absorb the swings between routine economic activity and the building of gigantic complexes solely from its own inhabitants, so America has created a class of construction gypsies. In a sense the WPPSS projects are an exception because they seem so endless that people like Lou and Dick can grow up in the region and almost accept as routine the prospects of continual work with no need to migrate. But perhaps a quarter of the WPPSS work force came only in order to build these plants, and when (and if) construction ends, they'll ride back into the sunset like the cowboy pipe fitter.

If the attitudes of craftsmen brought here by the Hanford boom differed from those of their peers elsewhere, it was due largely to the magnitude of their earnings and to their participation in a construction enterprise that never ended. When involved in normal building, they expected alternation of fatback and unemployment, but except when interrupted by periodic labor disputes, WPPSS work continued year after year without halt. For almost a decade now, no one had been able to, in one ironworker's words, "finish a project, then lay back and watch people using it."

But just as Hanford's founders never questioned what they produced in their reactors, so the young WPPSS workers remained at least publicly silent both about the situation that eroded their pride and commitment to craft, and about any misgivings regarding the atomic process itself. The pragmatic old hands believed that if the government ordered the plutonium plants they must be necessary. The young construction boomers viewed the massive presence of the new reactors as reason in itself for chasing the wealth that came with working on them. They used the sex and drugs and rock 'n' roll as a compensation to blot out any unease at jobs many admitted were useless. They rationalized contradictions by making the entire world part of an exquisite game whose codes were summed up by the maxim: "I fucked the company. The company fucked WPPSS. WPPSS fucked the government. And the government fucked me." And just as the old hands echoed, in their basement workshops, the Area's original task of invention, so the young boomers recapitulated in their private lives the WPPSS effort's waste and its things-will-get-done-whenever-they-get-done fatalism.

Gary was a WPPSS ironworker who lived downstairs from the Kennewick teacher with whom I stayed. One day he saw all the plants the teacher had hanging throughout his house and, deciding he'd like some himself, spent several hundred dollars filling the entire back of his pickup with ferns, African violets and coleus. When, six months later, they'd all died from lack of watering, Gary looked at the empty pots and said, "I think I'll go get some more plants."

Another time, when ironwork slowed briefly, a man came around recruiting from Libya. Gary was living with a woman at the time (he'd just gone through his second divorce by walking out and leaving his wife the house, the car and $50,000 in uncontested community property), but he came back from his successful interview, told his girlfriend he had to go and within two weeks was sending postcards saying "Sure is weird here. All these camels and all this oil. Well I'm here with the A-rabs getting my share."

One of Lou's pipe-fitter friends gave a party at the Pasco house he'd bought with his WPPSS earnings. The young workers drank from the keg of beer he'd provided and discussed how many miles per gallon their Porsches got. They took out $1500 concert-model Martin guitars and sang "The Ballad of Me and Bobby McGee" just like any veterans of the era when Janis Joplin made the Kris Kristofferson song into a road culture anthem. Inside a brown Dodge van parked outside, a half dozen young guys sprawled on a thick shag carpet in front of the portable TV which, along with a quadraphonic tape deck, had been built into the van by its owner. The show they watched, a "Saturday Night Live" sketch entitled The Pepsi Syndrome, described a nuclear accident triggered when someone spilled a Pepsi on the main control panel (a Seven-Up would have caused no harm, a character explained, "because it's the Un-Cola"). Lights blinked. Sirens sounded. Radioactive water spilled into an adjacent room. The operators called in Garrett Morris as a black cleaning woman and, assuring her it was just a routine job, asked if she'd do them a favor by going in and mopping up. The show ended with her a mutant giant and Jimmy Carter, having gone into the room and himself grown to fifteen feet, announcing, from a window high in the wall, that they were eloping.

The men in the van, all young WPPSS workers, loved the skit. "Red tag, red tag," they called out. "Where's Strand?" referring to the then director of WPPSS. "Must have been your weld, lazy fuck," one said to the other, because it was of course people like themselves who'd put the plant together. They repeated "Yeah, go mop it up," because grunts, not supervising honchos, always dealt with nuclear mess.

Later, at the Cosmo Angus, Dick, the New Wave carpenter, sat with his friends Karl and Craig by the corral-surrounded sunken dance floor drinking Black Russians and tripping on the mushrooms Greg had bought that afternoon. As a Joe Jackson record played from a background tape, Karl—a big bearded Swede who limped from a motorcycle accident Craig said later "left him with nothing but drugs and rock and roll to live for"—told us irately, "He stole that song! It's a rip-off of Elvis Costello's 'Watching the Detectives.' He can't do that!" Dick announced, to no one in particular, "I think I took too many mushrooms." Craig, who when not working at the Area, played gadfly in the letters column of the *Herald* and was working on a Tom Robbins-style novel, explained in a deadpan voice, "I was born in a log cabin twenty-three years ago in Portland, Oregon, and I always said 'Daddy, when I grow up I want to be a Hanford laborer.' "

When the band began blasting a mix of Stones, 60s rock and New Wave covers from a speaker one foot away, we decided to move to where we could hear ourselves talk. Walking the length of the long bar, we made ourselves comfortable in a small circular side room. Karl, who worked taking inventories of mechanical parts for WPPSS contractor J. A. Jones, explained that he made $10 an hour in the largely female Office and Professional Employees' Union compared to the $7 he'd get in town or the $14 or more with a high-pay outfit like the Teamsters: "But they've got us by the balls [because his union didn't have the clout of the skilled craftsworkers] so in the last round when everyone else got 15 percent more we had to settle for 9."

They ordered another round of Black Russians to go with the mushrooms (which Craig and Karl had also dropped) and the earlier beer. When the band's rhythm section fell slightly behind the beat, Karl announced, "The bass player just lost it," and left to watch a rock show on TV. Dick and Craig sat watching assorted young bucks and cowgirls swivel-hipping their way past us to the dance floor. A blond woman, who had given Karl a we-are-not-amused look when he began doing Monty

Python imitations was now sitting on the lap of a man in a fancy jean suit and fingering his pressed blue shirt collar. Dick remembered a boom in 1977 when they hired an entire group "on permit"—which meant they had only to say they were carpenters, go to work immediately, and then pass a written test three months later. "Everyone knows you can buy the tests," he said, "and here I was an apprentice telling supposed journeymen how to do the simplest dipshit tasks."

I remembered the same charges being made and documented in a 1979 *Tri-City Herald* article, in which a reporter paid $20 at a local lumberyard for the questions to six different tests that the union gave. Dick continued by recalling when he worked WNP 1 as a carpenter rigger, "which is the same as being a flagman, except that where any normal place would use walkie-talkies, I signaled my crane operators through telephone lines that were about as secure as if we were kids using a string and two tin cans." If the lines failed in mid-pick the operators would keep lowering their crane load until it hit something. "It was damn lucky no one was hurt the two times it happened, but we had to threaten to strike before anything got changed.

"I consider myself a craftsman," Dick continued. "Believe it or not, it took some time to learn what I know. I don't like standing around all the time. I don't like that I could be fired for moving some lights five feet instead of getting the electricians out of the shacks where they spend all of their time drinking coffee and staring at the three years' worth of *Playboy* and *Penthouse* foldouts on their walls. I don't like working in a place that makes no economic sense and where the government lets everyone pass the buck. I don't like that . . . Oh fuck, I left my dope in Karl's car and now there's nothing to smoke."

It was 12:30 by now, and the bar was heating up. We returned to the dance area to watch the men rest their hands gradually longer on the tight-jeaned asses of the Hanford secretaries and other office workers they pulled closer to their bodies. The women, hopping from lap to lap like children at a family gathering, began to shift from guarding their honor to purring and cuddling. The band played an Elvis Costello song and the music filled the floor.

Craig and Dick didn't dance, although Greg swung by with his eyes like two flying saucers, holding and whirling round a succession of women; letting one perch girlishly on his knee while he told her stories of romantic drug adventures. Craig and Dick ignored it all, though one cuddly little blond woman full of dear-me-cross-her-heart gestures flirted briefly with Craig and was labeled by him afterward "a real waterhead."

He brought the subject back to Hanford by suggesting jokingly that "we should build the plants like the Russians—round-the-clock shifts and no containment domes, just go ahead and lose a few lives." He thought the Three Mile Island operators were told so often meltdowns were a one in a million chance that they couldn't believe it and react properly when it actually did start to happen. He recalled first joining the laborers' union, trying to show off how hard he could work and being told by everybody, "Slow down, it all pays the same."

Greg spun by grinning, with yet another woman in tow. Dick—though the band wasn't terrible—began saying repeatedly, "They're lousy, they're just lousy," and yelling "You suck!" as loud as he could at each instrumental break. A cowboy-shirted fitter they knew and a dark-haired woman snorted up right at their table while Craig commented dryly, "Guess they're going to have a White Christmas," and I thought of a Hanford ironworker who used coke as his early morning wake-up each Friday and Monday. The band played the song "Cocaine" a few minutes later, and the entire bar yelled and shot their fists into the air on the refrain.

As the evening ended, the young construction hands headed out to a snow-covered parking lot. They jawed with their buddies and flirted in final lazy attempts to score with the few remaining single women. Then, when these last efforts failed, they gunned their cars one by one into the night.

6

Ozzie and Harriet Meet the Breeder Reactor

The lobby at WPPSS headquarters contains a computer game that estimates electricity consumption based on where one lives, how much one heats, cools and insulates, and on which household appliances one uses. Visiting there my first day at Hanford, I was punching in my data when the tape displaying the results became stuck. First one and then four more WPPSS employees gathered around to deal with the problem: slapping the machine as one would a broken TV set, looking behind and beneath it for shorted out wires, rocking and shaking it to listen for broken parts inside. But the more they attempted to deal with the problem, the more lights flashed in increasingly strange and disconcerting combinations and the more the would-be fixers conferred in tense, worried tones. I fantasized for a moment that this was the beginning of a meltdown: my own private China Syndrome beginning in these bureaucratic offices and burning until it arrived in neutron-emitting glory on the desks of the Washington D.C. movers and shakers who'd created the industry to begin with.

The computer at the Hanford Science Center—a simulator outlining national energy production and demand—worked flawlessly. Though it was generally used in group presentations, I had access to it in an anteroom just past the gumball machine which dispensed the "atomic marbles" that had been darkened through gamma ray exposure. I set the dials to draw on coal, oil, natural gas, atomic fission and hydroelectric power; divided the outputs, where there was a choice, between chemical and electrical energy; then used that energy to satisfy needs for cooling, heating and transportation, for industry, construction and agriculture, for residential use and long-term research and development. Other dials determined population growth and consumer demand per capita. The machine continued running year by year until resources (such as natural gas) ran out, and other outputs had to increase or demand to drop.

The simulator's standard setting, strangely resembling then-President Carter's national energy plan, left the nuclear dial cranked just short of the point where a light marked "waste disposal" would come on blinking yellow for a warning and red for danger, and pumped Research and Development as high as possible to create new energy technologies which would kick in twenty years down the line and overshadow all current power sources. Growth and consumption remained high. The machine ran forever with only minor adjustments.

When I tried eliminating nuclear energy I ran into trouble. I could turn that particular input to zero—the Battelle Corporation program cranked out options, not ideological viewpoints—but all nonfossil-fueled technologies were pushed far to the future in terms of potential usefulness. Conservation was equated with turning the consumption dial down, not using resources more efficiently. Existing alternative approaches like co-generation (the use of excess heat created as a by-product of industrial processes, which supplies 27 percent of Soviet and 10 percent of Swedish electrical needs), passive solar construction or solar hot-water heating were not even mentioned. I could have kept consumption high by burning so much coal that a red air pollution light would have come on to indicate the equivalent of emphysema for the entire nation, but that seemed an unwise choice. Instead I cut growth of population and of energy use down to zero, then limped along with no air conditioning, minimal heating and limited personal transportation until, after twenty years of hair-shirt privation, the "new energy sources" at last kicked in.

Like the simulator, the Science Center in general allowed for a variety of tastes. Pictures and biographies of atomic pioneers hung from the orange-girdered ceiling to serve as room dividers. I stacked blocks with a version of the remote manipulators which welded cesium by-products inside steel containers at the Hanford plants. I placed my hands in sealed glove boxes like those Amy and her co-workers used to handle plutonium. I punched out S-U-N to answer a computer crossword game's question regarding the earth's ultimate source of energy. Playing with twin Geiger counters, I learned a watch's radium dial was more radioactive than displayed samples of strontium 90, uranium ore and thorium oxide. In a constantly repeating Disney slide show, a character named Energy Ant ran a windmill, boiled water, mined coal and generally progressed from primitive inefficiency to a clean atomic tomorrow.

Given that Rockwell Corporation ran the center just as they ran the waste program, the laundry for contaminated clothing and the bus fleet that drove employees out to the Area each morning, the atmosphere of stalwart optimism shouldn't have surprised me. There were displays on solar, wind and geothermal power (accompanied by explanations that despite their potential they'd require billions of dollars and many years to be practical), and the library did contain critical works—*Environment* magazine, *Bulletin of the Atomic Scientists* and John Gofman and Arthur Tamplin's book, *Poisoned Power*, for example. But the library was primarily for researchers and scholars, and the mention of renewable technologies was so circumscribed with proclamations of difficulty that the effect was to anticipate and dismiss potential comparisons with the nuclear industry.

If Hanford's cultural institutions provide only tunnel vision, how can the Area's workers assess the safety or wisdom of the jobs they do each day? When I talked with Columbia High's current student body president, he explained their mushroom cloud symbol as "no different from any other. We're just proud to win a lot of games with it." Ads in the Richland yellow pages placed clear priorities between distraction and friendship by asking, "Why run to answer the phone and miss part of your favorite TV show? Why not enjoy the convenience of an additional phone in your living room?" The same Richland public library which recently installed a $17,000 electronic book theft detector fails to carry *Rolling Stone, Mother Jones, Ms.*, or any other periodicals indicating that Hanford kids grow up to be anything except split-level replicas of their parents.

Tri-Cities has four Bible bookstores, six regular outlets, and one porn shop. In the Bible stores, works such as Pat Boone's *Whatever Happened to Hell?*, Jeb Stuart Magruder's *From Power to Peace* and Anita Bryant and Bob Green's *At Any Cost* (which explained, "When the pie was thrown in Anita's face all Christians got that pie. When they attacked Anita they were attacking the body of Christ and His people") clearly offer little challenge to the Hanford workers' institutionally compliant attitudes.

On my first Hanford visit, John McPhee's *The Curve of Binding Energy* was the sole work I was able to find in any Tri-Cities bookstore which was at all uncomplimentary toward the nuclear industry. Although several outlets carried *The Mother Earth News* so the tinkerers could get ideas for heat pumps or even solar hot-water heaters, exploration of atomic controversy seemed to be confined to books with such titles as *The Health Hazards of Not Going Nuclear*.

When I returned a year later, the Richland store called The Book Place did have recent critical works by Helen Caldicott, Barry Commoner and Amory Lovins—and one of the proprietors said that during my previous visit they must simply have been out of generally stocked titles such as *We Almost Lost Detroit* and *Small Is Beautiful.* Although the counter held a new mass paperback edition of *The Health Hazards of Not Going Nuclear* and a pamphlet by a local engineer which gave the standard explanations of how living near a reactor exposed you to far less radiation than did taking a jet plane flight or living in high-altitude Denver, I wondered if some formerly "beyond the pale" critical voices had gained at least slightly greater legitimacy in the wake of Three Mile Island and WPPSS's massive financial morass.

When I asked the owner of the Kennewick Bookshop why—given the concentration of highly educated residents here (Richland is rumored to have the highest Ph.D. population in the country)—she carried no works of philosophy, history or social theory, she explained that there was no demand. Then a man waiting in line jumped in to elaborate, saying that books of that sort belonged to "an ivory tower world"—which he had encountered in college, but which had no application to the real one Hanford inhabitants were dealing with here.

This is not to say the Hanford environment is barren of intellectual challenges. With cold war paranoia having eased since the time when John Ozaroff threw his wife's Russian books in the river, Columbia High kids can now study that language so as to keep up with the latest technological advances. The Battelle-connected Center for Graduate Studies brings in noted lecturers and workshop leaders in a variety of fields. The plants themselves are often working laboratories for experimentation.

But unspoken limits unite both Hanford's official and its unofficial cultural institutions. At one frontier—that of nuclear production—researchers can push beyond all previous limits to explore processes barely conceived of forty years ago. But because this effort has demanded such a massive investment of money, time and hope—and because not questioning the work makes life far easier—all challenges to the atomic project are feared. The more the potential risks of nuclear technologies emerge, awesome and incompletely understood, the more the products and processes of the reactors must be made to seem as safe and mundane as those of any other industrial enterprise.

I had been forewarned about *Tri-City Herald* editor Glenn Lee's temper. Two Hearst Corporation executives had come in to discuss buying the *Herald* shortly after the Hearst-owned *Seattle Post-Intelligencer* had broken the story of the 1954 release from a Hanford processing plant of ruthenium-contaminated particles that were then taken up by air currents and deposited in a 400-mile radioactive trail from the Oregon border to Spokane. Greeting the men by holding up the *P-I* issue with the Hanford headline, Lee is said to have begun screaming, "You let your people print this? You'll buy this paper over my dead body," and ended up selling instead to the *Sacramento Bee*.

After repeated interview requests, Lee ended up giving me fifteen minutes and announcing, as he glared like a Marine Corps general, that he wouldn't tolerate "any goddamn leading questions."

"Hanford has no problems except those common to communities everywhere," he said, jabbing his finger in the air for emphasis. "People speak off the record in all government towns, not only here . . . No, that doesn't give my reporters problems—and I thought I told you already not to ask those kinds of silly questions."

The *Herald* has been compared by one of its reporters to "the *Dallas News* of the 50s," labeled "Voice of the Mid-Columbian Empire" by its ad and circulation departments, renamed, as previously mentioned, "Tricycle Herald" by nearly all Hanford workers aged mid-thirties or younger, and considered an unsound risk by Lee's insurance company—which made him drop his front page column after the seventh successful libel judgment against the paper. The yearly Progress Edition, the size of a Sunday *New York Times*, mixes extended versions of Nuclear Industrial Council press releases (the council's founding triumvirate consisted of Volpentest, Lee and the *Herald*'s publisher) with praises for the booming new fast food industry. Lead photos in the regular issues consistently show such images as car crashes and men giving mouth-to-mouth resuscitation to dead children. Wire service stories focus on teenagers freezing to death and glue sniffers whose parents donate their hearts to worthy recipients. Although the *Herald* does print news of specific nuclear industry problems, it presents them as but minor obstacles in an onward and upward march toward the future.

"Write this down," Glenn Lee said as we continued our interview. "We've killed 100,000 people on the highways. That's a human loss beyond calculation. We haven't killed a single person in a commercial reactor accident for the whole history of this industry, but you can either go forward with progress or back to caveman days like the no-growthers want. If you look at scientific facts, we're going to starve to death without additional food and energy in the next thirty-five years, and if that starts happening we'll have revolution."

When I asked Lee to elaborate on this last statement he told me, "Look, if I wanted to go into philosophy I'd have become a preacher. But you know a takeover is just what Russia wants. Not all the do-gooder operators get direct support. But how else do 75,000 people come to protest nuclear power and then leave their demonstration site so destroyed it takes a week to clean up the condoms and beer cans?"

Hanford culture today tends toward two extremes: The boomers ride their drug roller coasters, grab their sex on the run and seize all manner of other instant highs; the sedate old hands jettison all unruly desires for a civilized present and hopes of a techno-paradise future. The boomers are interchangeable and—although their work requires skill and often meticulous effort—ultimately replaceable; the knowledge that the old hands have of the projects they built cannot be duplicated through instruction courses, blueprints or procedures. Though the boomers still carry rebellious impulses as legacies of the 1960s, their transience denies them political voice. And the old hands, who have a stake in the surroundings that have now become their home, value restraint and order far too much to ever challenge official institutions.

In a sense the old hands still live in the world Lou described: They resemble characters in a June Allyson film that reruns endlessly while the implications of an ever-more complex technical future is obscured; they remain frozen in the time, 1961 to be exact, when they became an official All-American Town.

One can see the best of that era mirrored in the old Hudsons, Chevys and Studebakers that sit in many Richland driveways, their Brancusi curves lovingly restored by the tinkerers to make the town a Twilight Zone warp to the past.

The fact that Richland is white, clean and well-swept, and that it has no street culture of Italian fish peddlers or Jewish garment merchants or Puerto Rican salsa kids, merely indicated that is is not metropolitan and that its immigrants have long since left behind all old world roots. But the town also has little of the clapboard-shade tree-swimming hole-and grange hall ambience of even old Pasco. Its inhabitants have spurned equally unseemly urban ferment and the archaic style of the surrounding rural culture.

It's not just the old hands' reticence that has created the Tri-Cities Moral Majority; the local new right coalition includes Hanford workers of all ages, as well as a large contingent of Mormons (10 percent of Tri-Cities' population) who came to work not just at the Area, but also in agricultural enterprises like the corn, wheat and potato farms begun after U & I Corporation moved its national headquarters from Salt Lake City to Kennewick and bought and irrigated over 100,000 acres in the Horse Heavens. But, believing that all non-procreative sex is the devil's work, the sentinels of decency joined together here to push through a Richland ordinance which—until the city council decided to hold it in abeyance pending court challenges of a similar one in Akron, Ohio—would have made abortions virtually impossible by requiring attending physicians to describe in minute detail the physical features of the fetus, then to ask the mother whether she wished it buried or cremated. Mormons on the Kennewick School Board wrote sex education out of the approved curriculum—threatening teachers with dismissal if they mentioned sexuality even in Modern Health courses—and, after banning donations of paperbacks to school libraries (which meant that potentially controversial books had to go through formal approval channels), pulled out works by such authors as Kurt Vonnegut, whom they considered morally objectionable. Local ministers even pressured Tri-Cities theaters into ignoring extrordinarily heavy requests to show the Monty Python film, *Life of Brian*, and when a friend of Virginia Dumont's held a National Abortion Rights Action League fundraiser, the guests were greeted by a prayer vigil, complete with lit candles, going on at the edge of her driveway.

Virginia herself witnessed one confrontation at a local church where Planned Parenthood was showing sex education films for parents and for teenagers, with open discussions to follow. Brandishing their numerous children as proof of their righteousness, the moral legionnaires screamed and shook their fists in the face of the woman who was moderating the discussion, demanding the films be withdrawn. "It was terrifying," Virginia said, telling her story at the bridge club. "Their eyes shone like Savonarola because they knew what was right and wrong; they knew that sex outside of marriage was sinful and that abortion was murder. We couldn't even get them to hear what we were saying because they all had their own direct lines to God."

When Virginia spoke, Barbara and Sarah nodded sympathetically, but the others sat stolid and expressionless. All cooperated in moving the conversation quickly on to a less threatening subject. Most believed, as Sue said at another session, that they did fine without abortions when they were young, and the assaults on sexual choice rested with other controversies they saw no need to deal with. Although some of the men, among them Sam Beerman, believed strongly in civil liberties and rights—and others such as Clark Reitnauer even believed victimless crimes should not be prohibited—they were too busy with their work to pay much attention. And because the boomers weren't planning to stay, they made no squawk about the push to legislate sexual morality or even about a law, the first such in the state, that was pushed through by an antidrug organization called PRIDE to ban all paraphernalia.

When the Kennewick High School paper ran a student's article on fossil hunting in the Horse Heavens, local Seventh Day

Adventists called to explain that the article lied because God put the fossil bones in the hills at the same time as He created man and all the animals. It seemed strange at first that members of an antievolutionist church would hold major positions in an industry as scientifically sophisticated as the nuclear one. But Hanford fundamentalism, while threatened by Darwinism, made a very pragmatic accommodation to atomic power, justifying their position through constructs like those of a theologian who explained in the American Nuclear Society magazine that if you traveled between the stars, God would show you many nuclear reactions but few other sources of energy, so in no way could atomic power be viewed as un-Christian.

At Kennewick's United Pentecostal Church, I met a gray-suited twenty-one-year-old WPPSS printer named Duane. He'd brought his wife, whom he'd married right out of high school and who now sat gawky in a flowered dress, listening teary-eyed to the services. He'd brought his sixteen-month-old daughter, who wore a pink dress, pink rubber pants, pink socks and pink shoes, and who kept tottering over to me with a blue alphabet block in her mouth. Duane played drums in the church band along with a used car salesman on trumpet, a Rockwell welder on guitar and a Culligan Man bottled water deliverer on trombone.

The old wooden church was decorated with orange and yellow carpeting that looked like it might not last until the imminently projected Second Coming. The preacher, dressed in sober black, intoned: "Lord, we're just your lost people. Give us the Holy Ghost." The congregants prayed in the aisles, on the floor and below the pulpit—asking for forgiveness and aid in selling more church fund peanut brittle. A taped, slide-accompanied sermon by the former Annapolis officer who founded the sect showed how a hippie couple—"he was a motorcycle leader and she was involved in witchcraft"—became gloriously united in Christ.

When Duane asked me if I believed in the end of the world and I said I didn't, he explained, "But it is coming!" Then, calling this "the times of tribulation," he cited the instance of wars, rumors of wars, divorce, rising homosexuality ("you know, that's what they named Sodom and Gomorrah after"), and the "mark of the beast," which Revelations had foretold and which already had appeared in the form of the Universal Price Code. For the meantime, because he had reaped many blessings from the Lord, he expected personal hardship up ahead.

As the band was getting ready to play again and Duane to join them, I asked whether he thought nuclear power was the work of God or of Satan. "It's godly," he answered, "because we need it to provide power for electricity, and God gave us intelligence so we could make things to use."

Since Duane had said earlier that most people looked out only for themselves, I asked if more goods wouldn't just make everyone more selfish.

"No," he answered, "because it's whether you accept God or not that counts."

The presence of Pentacostals and other millennial religious groups is certainly substantial at Hanford, but their beliefs are also far from the norms of most old hands. For the majority here, a happy medium is embodied in such churches as the Central United Protestant congregation attended by most of the bridge club members: tasteful houses of worship capable of meshing their pronouncements smoothly with the given assumptions of Hanford life; distanced equally from the fundamentalists and from such men as William Sloane Coffin, Martin Luther King or Dietrich Bonhoeffer who have used religious values to challenge institutions of power.

Believing his congregation represents community and that the more people he can draw in, the more human bonds will be built, CUP pastor Joe Harding advertises "a positive faith for today" in omnipresent radio commercials and endlessly proselytizes for new members in an effort that has made his one of the fastest growing Methodist congregations in the country. Church members donate canned goods to be sent to the poor, launch multicolored helium balloons on special holidays and attend guitar-accompanied early morning folk services. Though Harding strongly supported Nixon and was disappointed at his downfall, he considers himself a liberal, criticizing the Moral Majority when asked what he thinks about their program, allocating church money to the Richland Battered Women's Shelter, and even displaying works of Albert Camus and of radical Brazilian educator Paulo Freire on his office bookshelves.

We met in his pleasant office which was decorated with a weath of congenial accoutrements including—to symbolize interfaith cooperation—a menorah. Harding explained that, in contrast with the apocalyptic believers for whom—since Christ was returning imminently—environmental issues didn't matter, he thought churches "should emphasize stewardship of the earth." But while he didn't like making more bombs, he also wasn't a pacifist and saw no problem with the plutonium economy that breeder technology would usher in. Nuclear power was necessary to save oil for Third World industry and food production. Although he agreed that perhaps he should use some sermon to discuss the morality of breaking silence when operations violated safe procedures, Harding said that criticizing the atomic industry here "would be equivalent to taking a stand against smoking in Durham."

So aside from bringing in a critical National Council of Churches official (in part so the Hanford engineers could convert him), Harding had never discussed the ethics of what Hanford created. "If the nuclear industry is safe," he concluded, "it will be important that churches like ours continually urge that its benefits affirm ethical values. And even if there turns out to be a danger, at least we'll have been providing a balance by upholding the sacredness of human life."

On a later trip, I attended services in Central United's new modern building. Harding began by asking the congregants to "greet your friends and neighbors." They complied, turning around in their pews to shake hands and to wish everyone in reach a good morning. A choir sang into a battery of microphones hung in perfect acoustical array. Harding baptized a red-faced infant, exclaiming, "He's going to be a wonderful young man!"

When the sermon—about how God's grace was like catching up to a snowplow while driving through a lonely winter pass—was over, I talked to a woman who said that when the FFTF experimental breeder came on line, someone added "and bless the FFTF" to the list of daily prayers. Two months after my visit a form letter thanked me for my joining them, hoped I'd "experienced the friendliness and enthusiasm of this wonderful congregation," and concluded that my presence helped them "feel alive and growing and grateful to God."

The monks of the Society for Creative Anachronism had a more secular task: to uphold chivalrous traditions by re-enacting medieval pageants. I discovered them at Richland's annual arts and crafts fair: four women ranging in age from their late twenties to early thirties, some wearing monks' robes and others in the white-bodiced dresses of bawdy Elizabethan serving wenches. They sat in front of a tent decorated with heraldic shields and yellow fleur-de-lys tapestries. They sang choral rounds of "Rose, rose, will I ever see thee wed" and "Parsley, sage, rosemary and thyme." They called each other "clumperton" (clown in Elizabethan English) and "flibbertigibbet" (gossipy old lady); their name for Hanford, according to one woman named Victoria, was "Wastekeepshire."

She and her friends, Victoria explained, constituted a local chapter of a national medievalist organization. They'd put together an archers' guild and an annual "Twelfth Night" October ball. The Queen Elizabeth costume she'd made with one thousand hand-sewn imitation seed pearls had won first prize at a fair in Agoura, California. She hoped I'd come to see their sword fight tomorrow.

When I asked Victoria if she worked at the Area, she replied,

"I am an artist," and showed me line drawings of long-legged women with sparrows circling around their heads. She was supported by her husband, whose job involved estimating design costs for Vitro Corporation, the Hanford architecture and engineering contractor. They moved here from California eight months ago, after he'd answered a Department of Energy ad in a San Bernardino paper. She had become involved in the Anachronists after a Vitro engineer started a Tri-Cities chapter. As for the environment here, Victoria labeled it "mundane" and quickly changed the subject to the beliefs of Thomas Aquinas.

When I asked her fellow-Anachronist Janet about the connection between Hanford's futuristic technology and dreams of a return to medieval life, she stated emphatically, "There are none." Because she was six months pregnant, she admitted the possibility of radiation exposure did make her worry a bit. "But Richard, my husband, doesn't," she said. "I joked once that maybe he should put a dosimeter badge on his trouser fly. But he just repeats these dumb slogans like 'A little nukie never hurt anyone,' then grins at me as if he's proving something. He was never like that before we came. I'm not really scared but it bothers me."

A few days later I stood with Victoria outside her new house just south of Kennewick, at the edge of the Horse Heavens. Five years back, she'd have settled down in Richland. But with nearly all space there occupied, her street with its six-month-old homes marked the far border of Kennewick's boomtown development. Victoria was dressed conventionally now, in a peach blouse and maroon slacks. We looked to the north and to the west past Richland, where the WPPSS plants rose up, 25 miles away, appearing little different from the clusters of shopping centers and tract homes which formed leap-frogging quilts of development across the former orchard sites below us. Victoria told me of her "very antinuclear friend who just wishes I weren't so laissez-faire about everything. That man up there," she pointed to a Tudor castle perched, along with a half dozen more conventional houses, on the ridge above us, "perhaps he built that marvelous structure just to escape what's here. Maybe my neighbor is doing the same thing with his hand-built log cabin that's paid for, of course, with money he makes at Hanford. And maybe when I spent a month sewing pearls on that idiot dress, it was mainly to imagine myself in a time when everything was courtly and nice and I wouldn't have to deal with this world.

"Sometimes I look at that breeder out there and feel like we're Christians being fed to the lions. If we're bad and break the laws it can go boom. But even if we're good and do everything right, is it safe?" She pointed again, this time to the jagged white outline of a condominium development named Grandview Village. "So while we're making something that can blow us clean off the earth, they give us mock English towns and tell us everything's fine."

To a degree the culture here has evolved with the times. Although when Virginia returned to teaching, her friends acted as if she'd betrayed her proper wifely calling, within a few years several of them were working as well. Frances ended up teaching grade school, Doreen nursery school, and Ethel music. Barbara Giroux became a lab technician at Kadlec Hospital. Though not all were totally enthused about what they did, their kids were now grown and their husbands remained busy with their official and unofficial tinkering; as Barbara told Virginia once, "I don't like working so much, but I can only do so much housekeeping and spend so much time in bridge club. I'm no longer sure what I'd do if I didn't."

At times they were more tolerant socially. When Henrietta Beerman held an elaborate reception to celebrate the marriage of one of her daughters, she announced to Virginia—half as if to say "I guess we've all changed," half to show how daring she was—that she'd even invited three couples who were living together. A few old hands, though not many, shared Lester's feeling that "it doesn't kill me to know my kids have smoked marijuana." Even though everyone had disapproved of what

Virginia called "all the Haight-Ashbury flower children," she said they were "the beginning of a process where we stopped worrying on every occasion about making sure we had on the right little hats, white gloves and patent leather pumps." Bridge club members now supported with modest donations Virginia's Planned Parenthood efforts. When Washington held a 1972 referendum on a State Equal Rights Amendment, Tri-Cities support played a key role in getting the measure passed. A Richland audience even booed pro-nuclear spokesman Ralph Lapp for suggesting, in a debate with Ralph Nader and a female scientist from western Washington, that the reason women tended to oppose atomic power was their emotionalism. But when a Phil Donahue show touched off a bridge club argument on whether or not gays really were "just regular people," one of the women immediately changed the subject by asking enthusiastically, "Doesn't Donahue have interesting programs?" The group still shied away from all controversial subjects as if they were snakes coiled on the table, ready to strike.

During another visit, I asked Virginia her feelings regarding nuclear technology. She recalled living in the Mojave desert and taking for granted the heat which hung around tight as a choke collar, but which one simply had to accept; "then putting in our first air conditioner, opening the door to a cool room and thinking 'this makes living here worthwhile.' I had a Formica kitchen counter which wiped off easily, didn't stain and clearly made food preparation far easier. My niece installed a wooden one she said was 'more natural,' but it splintered and all the stains sunk in. Maybe the people in my generation have difficulty judging inventions we thought would deliver us everything. But I think many who came later grew up taking them so much for granted that you also have trouble sorting out which makes sense and which should be discarded.

"For most of my life I didn't question the desirability or safety of nuclear power. Though I've always been kind of a nonconformist, I treated it, just the way everyone else here did, as God and Apple Pie—and except for the waste problem, it still seems they should be able to do it safely. The past few years I did raise the few questions about waste disposal that, as I told you, Lester pretty casually dismissed. But when you have questions, there are no options here except asking the men who go out to work with the stuff every day. If I did begin to publicly criticize, I know I'd be crucified far worse than for any of the feminist stuff. And for all our high technology and friendly community, this is still a small town whose social groups can be awfully harsh if you break the basic rules. For feminism I'm willing to take a stand, because I have a strong enough feeling that what I'm saying is right and just and must be said. But on the nuclear issue I've never had any impulse to seriously question. And in that respect, maybe I am a true member of the bridge club."

7

Boomtown Cowboys II: Thrown from the Saddle

Hanford now is an All-American layer cake whose different cultures barely touch each other at the edges. Although one can find boomers with Harleys and Dobermans on some of Richland's well-manicured and well-mannered streets, the town remains primarily the refuge of the original atomic pioneers, of the higher-level managers and technical people living in the new northern developments, and of some younger and less affluent Area workers to the south. Because the Richland cops, who still consider themselves almost a national security police force, are quick to crack down on any loud, disorderly or undignified activity, the quest for sex and drugs and rock 'n' roll stays far more discreet here than in Kennewick or Pasco. The old hands still see little reason to visit the other two towns except to buy cars or car parts in Kennewick or shop at the neutral ground of Columbia Center. And as Virginia said, they view the boomers "simply as 'those people,' the others. While we suppose their presence is necessary, there's enough physical room here so we can draw our wagons into a circle and ignore them."

The boom catalyzed by the FFTF and the three Hanford WPPSS plants arrived in a hyperpaced rush through the atomic-age horse trading of men like Sam Volpentest. But in a sense the speed of its appearance merely echoed that with which the original Hanford project was slapped down on the desert like a slab of bacon on a short-order grill. Aside from such essential attributes as isolation and access to plentiful water and power, the nature of the surrounding environment was unimportant either to the Manhattan Project or to the men who moved here to build the reactors. Although residents swam in the Columbia and hiked in the surrounding desert and nearby mountains, the mission of invention was too important for many to waste time or energy contemplating sand and desert sagebrush. At the Area, radioactive hazard and security restrictions discouraged leisurely desert strolls. The community in general treated the land it settled on as obstacle terrain to be mastered.

The old hands endured the termination winds, the sand which filtered in beneath their doors and the grit—in part stirred up by the massive construction effort—which found its way into everything they owned. They helped save the town in 1948 when the Columbia flooded and Hanford workers, together with army troops, worked round the clock to throw together an earthen dike they named the "miracle mile." They sometimes labored as lovingly on their box houses as they did at the machine systems of the Area.

But the atomic pioneers were also limited by their context. The houses, all turned out of a few quick-set molds, could not be adapted to the land on which they sat. While in California's Imperial Valley thick pastel walls of stucco or adobe kept rooms in many of the old homes relatively cool despite equivalent 110-degree summer temperatures, here insulation was unheard of, and the homes were appropriate not to an arid desert but to Nebraska prairies. Even the later additions often clashed with the houses' basic structure—like one brown stone facade that was placed over flat white boards like a lizard-skin graft on the wing of a dove.

Although the old Richland streets remain demure and tame, Hanford's recent boom has brought a new era of slapped-on-the-landscape construction to the most recently opened northern section. The style now is cedar condominiums with signs saying "If You Lived Here, You'd Be Home Now," and block-long courts of motel-style apartments, each separated by white concrete fences, like giant shower stalls. Names such as the Illahee attempt to convert the desert heat to the soft sea ambience of Polynesia. Other clusters fill in the empty spaces as if the buildings were so many dice rolled scattershot across the landscape.

But Richland's new growth is still bedroom growth. The town, as it always has, lacks any except the very lightest nonnuclear industry and, in keeping with its genteel reputation, does not presently have even a single auto dealer. The needs for less expensive housing and less classy commerce are taken care of by Kennewick, whose expansion has been fueled largely by WPPSS boomers who view Richland still as the Ozzie and Harriet land to be skirted on the bypass highway while heading to and from work each day, and by people like Victoria and her husband who hold suburban family dreams but cannot afford the available Richland housing.

To get to Kennewick I drove, of course, for, as a creature of the postwar era, Tri-Cities is unquestionably a product of Detroit logic. Every few years someone runs a ballot referendum to provide the towns with bus service beyond the single run from the North Richland depot to Hanford, and each time the measure fails. Every morning and every evening traffic therefore piles up from the edge of the WPPSS sites to the Kennewick exit off the highway connecting the towns; people choke on their own exhausts as the lines of cars and pickups ripple through the desert.

But I was going in off-hours, driving by the old trailer colonies in the backwater strip between the Columbia and Yakima rivers called "the Richland Y," past Columbia Park with its pedestal-mounted model of an early Atomic Cup hydroplane, and past the turnoff to Columbia Center. I drove as fast as I pleased, like all the young kids who swept away difficulties in the rush of heat and motion that their Porsches and Trans Ams were ideally suited for. The environments I passed dropped away to a blur, and the radio picked up the down-and-drunk country road songs—coming across the desert on a nighttime skip-jump—of KFI-Los Angeles and of KMPC in Merle Haggard's hometown of Bakersfield.

Perhaps that was all that was required: the right vehicle in which to speed just ahead of whatever wave of uncertainty threatened to crash just behind you.

I left the freeway, pulled onto Kennewick's Columbia Drive and entered a six-block strip that began with a used car lot and a sign advertising the "Off the Hi-Way Motel and Trailer Park." It continued with three gas stations, a U-haul rental place, Baydo's Trailer Sales, two more used car lots, Michelin Tires and dealers selling Pontiacs, Cadillacs and Mazdas. It concluded with a lot full of $10,000 Toyota four-wheel drives, two more displaying assorted used junkers, the Drive-In Pet Hospital and a power boat dealer, Sundown Speed & Marine.

The Tri-Cities developed in an era when it seemed that pure mobility might allow us to leave behind archaic problems of poverty, class strife and ethnic and racial friction. We could zoom from splendor to splendor in machines which—reflecting our ingenuity, perseverence and strength—were shaped with rocket nose bulges, hydroplane fins and gun-barrel taillights. The cars' names—Ford's Custom 300, Dodge's Royal V-8 and the Olds Super 88 that blues singer James Cotton soon renamed "my Rockett 88"—assumed engine size or wheel base figures were in themselves so inherently seductive that they needed no further amplification. Later Polaras, Futuras and Comets became Detroit's personal answer to Sputnik's space-age challenge; then Mustangs, Impalas and Firebirds sped across open plains to celebrate a Great Society's bountiful future. Now, in a time when Americans—including many Hanford engineers—are settling for the Rabbits, Colts and Pintos appropriate to an age of lowered expectations, the atomic boom riders remain megapowered and overdriven as ever.

The result of this is a culture as thrown together as the atmosphere of a bar near the Columbia which melds portraits of English country squires in knee breeches and waitresses wearing medieval ruffled blouses and creampuff hats with studs swaggering in wearing designer jeans and Tony Lama boots. The houses in Panoramic Heights, an exclusive Kennewick development overlooking the three towns and the Columbia, balance Richland's melba-toast blandness by combining every imaginable architectural style into a jumble of peak roofed chalets, red-arched Spanish haciendas, neo-Pizza Huts with low shingle skirts and gingerbread fantasies resembling something Jay Gatsby might have built to upgrade the New Orleans French Quarter.

The crowning jewel was to be another Tudor castle, built to incorporate some item from every fifth page of the standard design handbook, and its highlights including a dungeon basement with drawbridge entrance, a French provincial den and five sunken bathrooms.

Most houses here, explained the young architect who was guiding me through the development, were not nearly as monstrous as this. But they were still "a grab bag where people think 'this looks nice and that looks nice,' but never understand how different parts should fit together.

"Look at those cedar shingles," she said, pointing to a half dozen houses directly below us. "People use them because they're supposed to be northwestern or something. But while they'd last forever in a damp, shaded environment like that of Puget Sound, they'll start to go inside of five years in this dry desert heat. Though there are plenty of inexpensive and durable alternatives like clay roofing tiles, they slap these up and live with the consequences."

A year after my initial visit, I called Dave, the Park Lane pipe fitter—and got a telephone operator who asked if I wanted his new number in Oregon. FOR RENT signs everywhere marked the rise of apartment vacancy rates—1 percent to 1½ percent the year before—to between 16 percent and 17 percent by November of 1980. Pasco's largest pawnshop hung a sign announcing NO LOANS.

What had hit Tri-Cities was five months of what everyone called a WPPSS strike, although more properly it was a lockout.

By halting construction, it threw out of work most of the 9,000 men and women employed by WPPSS contractors, as well as by the shops, restaurants and other local businesses that served them. Approximately 2,500 wage earners left town altogether and perhaps another 1,000 took interim jobs elsewhere while leaving their families behind. Local classifieds overflowed with offers to sell Porsches, four-wheel drives and gun collections.

The dispute began when contracts for the operating engineers (who ran the cranes) and for the carpenters came due for renegotiation. Most Hanford unions had previously dealt with the Spokane-based Associated General Contractors in deciding on wages, benefits and conditions for varying types of work, but under unofficial WPPSS pressure AGC refused to deal with them. The new bargaining unit would be the on-site Hanford Contractors' Association, and the companies building the three plants refused to let the union members work without an HCA-negotiated contract. Without going into all the ways AGC and HCA differed, HCA was on-site, under more direct WPPSS influence and better situated to resist union demands in both the endless jurisdictional disputes and day-to-day work procedures, as well as possibly in terms of wages and benefits. Since without the carpenters' frames and the operating engineers' hooks almost no work could be done, and since at varying points other unions joined the dispute, construction essentially stopped from June 1980 until November, at a cost—since administrative overhead, interest on already issued bonds and general inflation all continued—of an additional $1.3 billion for Northwest ratepayers. Following threats of federal intervention, the unions finally settled with HCA, gaining 15 to 18 percent pay increases but surrendering a preferred bargaining situation and the right of their members to simply walk out without first following specific procedures involving both the union business agents and the contractors involved. If WPPSS took a short-term loss on the increased costs, they also clamped down on a labor situation considered to be the most unruly of any nuclear site in the country, and by doing so eased the way for all future atomic pork barrel for Hanford.

When excavation for WNP 2 began in 1972, the reactor was intended to go on line in 1977, as the earliest of the plants, and to cost $397 million. It is now targeted for 1984, with projected costs of more than $3 billion, and is claimed to be 85 percent complete by WPPSS spokespeople. But between November 1979 and February 1980, Quality Control inspectors went to the Nuclear Regulatory Commission with evidence of defects in sections of what's called the Sacrificial Shield Wall—a concrete shell, sandwiched between the reactor pressure vessel and the containment building, which protects workers and equipment from radiation and supports the reactor against external stresses such as earthquakes. Though WPPSS personnel certified as correct the work done on the wall by the Seattle-based firm of Leckenby & Co., it turned out that the wall's upper and lower sections had been improperly welded together. The defects in both this wall and in the whip restraints that held pipes in place in case they ruptured had been pointed out by previous QC inspectors but ignored; so the NRC fined WPPSS $61,000 and made it redo the wall (WPPSS is now suing Leckenby for $120 million). Moreover, inspectors have discovered fundamental problems in the plant's 1100-megawatt General Electric boiling water reactor, which—due to a combination of poor design and the shoddy work on the sac shield wall—has a tendency to vibrate under certain circumstances, causing major stress to all surrounding components. If these difficulties aren't solved, the YES WE CAN slogan WPPSS prints on its press packets may become in Plant 2's case almost a speculative prayer.

The other WPPSS reactors have had nearly equivalent problems with delays, poor equipment installation and faults in both direct construction and QA and QC inspection. With direct costs estimates for all five having risen from the original $4 billion to current, but certainly not guaranteed, projections of $24 billion, workers are beginning to call the enterprise "The

World Bank." A study commissioned by a dissident Washington legislator tallied expenses of fuel, operation and maintenance, insurance, taxes, routine repairs and waste disposal, over the thirty-five years WPPSS intends the reactors to last, added the costs of interest on the bonds and of plant decommissioning, and after factoring in an 8 percent annual inflation rate came up with a total cost to Northwest ratepayers of over $200 billion. It's true that, as do all utilities west of the Rockies, WPPSS hopes to make money selling power to what they assume is an unlimited California market. And their spokespeople, disputing certain aspects of the estimate, claim the figure is really only $110 billion. But costs have soared so high that, as new WPPSS director Bob Ferguson commented, if the projects don't get completed and brought on line, "it could mean the end of the nuclear industry in the U.S."

Why—leaving aside all controversies over whether atomic power is salvation or doom—has WPPSS been unable simply to build and bring into operation these projects begun almost ten years ago? Craig, the writer and laborer I met at the Cosmo Angus bar, came up with a mythical WPPSS board game which he described in a letter to the *Herald*.

The game, Craig began, is based "on Monopoly, Easy Money and (unfortunately) on the game of Life." Players, pretending to be members of the WPPSS board of directors, receive unlimited expense accounts and unlimited taxpayer-financed credit lines. Since the winner is the one who spends the most, any display of competence, even if it's accidental, is severely punished."

To illustrate the game, Craig described a hypothetical contest between himself and John Doe. Doe begins, lands on "dinner and drinks with Senator Magnuson" (Craig created the game just before the 1980 election), pays just $600 and curses his bad luck. Craig hires Spencer Bishop (a prolific writer of cranky letters to the *Herald*) as a public relations consultant, spends $50,000 and announces rate increases. But Doe strikes it big, landing on: "NRC forces redesign of containment vessel.

Cost $230 million. Announce rate increases and move back three spaces . . ."

"There are ten NRC spaces," Craig's letter went on to explain, "the least expensive of which costs $27.5 million. All NRC spaces send the player three steps back. With any luck at all, the NRC will halt your progress entirely, costing you billions of dollars."

Finally, Craig draws a "Distribute the Blame" card that gives him the option of blaming a rate increase on construction unions and accepting a $5,000 raise for himself; blaming the NRC and accepting a $10,000 raise; or blaming a depressed bond market and accepting $15,000: "In classic WPPSS form I blame all three and accept $30,000."

The game captured accurately the range of explanations for WPPSS failures. The official WPPSS press kit cites changed regulatory requirements as causing 50 percent of all delays and cost escalations, along with strikes and schedule extensions (15 percent) inflation (30 percent), increases in the price of nuclear fuel (4 percent) and other authorized costs (1 percent)—all totaling a neat 100 percent explanation. The inspectors do come in, as Craig said, to make you go back three spaces and sock you with $10 million, $20 million or $200 million in mandatory design changes, and in the words of Benton/Franklin County Labor Council head Harry Alden, they offer the WPPSS engineers "a constantly moving target."

But given that the NRC often does not apply new safety regulations to plants already built, the prime chance to confront freshly acknowledged problems is in the building stage. When the construction stretches out for ten years and does so during a period when vulnerabilities in the atomic cycle are constantly being challenged, the integration of necessary modifications inevitably delays reactor completion.

That these postponements and overruns may well be endemic to the nuclear industry is evidenced by similar problems being faced by sites under construction around the country. Even the now famed Three Mile Island Unit #2 whose accident caused over $20 million in design changes to the originally identical reactor, WNP 1, jumped in cost from a 1969 projection of $310 million to $700 million by the time it went on line in 1978. But the WPPSS fiscal crisis is unprecedented for ventures of its kind.

Most Hanford old hands blame first the regulators and secondarily labor. A physicist I spoke with compared the pipe fitters and members of other high-rolling union locals to nineteenth-century robber barons. The availability of endless work for high pay does lead craftsmen here to desire as long a stay as possible at the golden trough, as well as the maximum individual share, even if it means combating other union members to get it. The generation which makes up the bulk of WPPSS workers (as well as many grunts in regular Hanford operations) in fact does not have the same vested loyalty to the nuclear endeavor as did their parents, who built the original plants. Since the revolutions of the 60s have, at least for the moment, failed, and since no one offers vast socially useful projects with decent pay, why not, as pipe fitter Dave said, take the money?

But the construction craftsmen here are not much less productive than those working other jobs for little love all across America. They proved this during the lockout when contractor after contractor testified at State Senate hearings that the problems of WPPSS were due not to the labor force but to the shoddy management and engineering which left high-paid workers standing around with nothing to do.

Whether or not nuclear plants can be built and maintained safely and competitively priced (with all subsidies stripped away) compared to other forms of energy production or conservation, WPPSS management clearly aggravated existing problems. Because the Net Billing assured that they'd always be able to float more bonds, they had no financial restrictions. Because the contracts ended up being cost-plus with no effective limitations, the more the contractors spent, the more they made—and nearly all overran their original estimates by one hundred, two hundred or even three hundred million dollars. Because no integrated schedules coordinated the thirty-odd companies on site at any given time, systems continued to be installed prematurely, then pulled because other key components would not fit. Contractors blocked each other's progress by setting up construction in areas where cranes and materials had to be brought through to work on other sectors. Crews stood idle, unable to help others performing identical tasks, but whose paychecks happened to be signed by different employers. The lack of coordination applied also to understanding the effect of regulations. As Labor Council head Alden said, they sent fitters out to do welding and carpenters to build the requisite scaffolding, then discovered "that the weld package—the procedural approval documents—hadn't been done, and since if you went ahead anyway you brought the NRC on your back, you ended up with a bunch of $150-a-day men standing round pitching pennies."

Ferguson, who took over in September 1980 to replace a man who'd been promoted from one trouble-plagued WPPSS project to another, recognized some of these problems in admitting: "The contractors don't realize that nuclear plants are different kinds of projects." He even brought in Bechtel Corporation to take responsibility for major preoperational stages (graffiti soon went up labeling one overzealous foreman "Rectal Roy, the Bechtel Boy"), which meant largely completing the job WPPSS, its managers and its multitude of contractors had been unable to do on their own.

Due to the Net Billing Agreements and further federal backing from a recently passed Northwest Power Bill, Ferguson and his deputy Alex Squire have still been able to finance construction through the endless succession of sales that have made WPPSS the largest municipal bond issuer in the country. Traveling east once or twice a month, they've held special luncheons with American Express investors at New York's Downtown Athletic Club, with Hartford insurance executives, with portfolio managers at Boston's Parker House, and with various

Midwestern financiers at the Chicago Club—all to draw further and further liens on the future of Northwest ratepayers. And with the projects still facing the legacies of previous mismanagement and of atomic technology's unresolved problems, the end of the game, as Craig would say, remains out of sight.

A year after our first visit, Lou was no less cynical about the Area than before—repeating "Problems? Have another billion," as if it were the official Hanford motto, describing how he was being paid $32,000 a year largely to reduplicate reports lost through general sloppiness, and telling me I could pick out the new engineers who came in with Boeing's construction subsidiary "because they all wear plaid pants." Yet his flip remarks died more quickly than before, and he seemed depressed as he explained how WPPSS could have used the lockout to patch up some of their problems, but instead lost skilled workers who would never return "and spent the whole time formulating reports, setting up task forces and playing games.

"Most people, myself included," Lou continued, "do construction because we like starting with a hole in the ground and working ourselves out of a job. This goes on forever without building anything. This isn't construction."

"Neither GE nor Du Pont were known for being pro-union," Labor Council head Harry Alden told me in describing the blue collar world of Hanford's early days. "That was back when Ronald Reagan was still giving antilabor talks for GE on the rubber chicken circuit, and we had to sue them under the Bacon-Davis Act because they were a government project paying less than prevailing scale. They called everything minor construction so they could pay maintenance wages, say $10 a day back then, instead of the regular construction rate of $18. We beat them alone, without any help from our national union." We, in this case, referred to Plumbers' and Steam Fitters' Local 598, of which Alden was a member and later president; the local fought the company on radiation standards as well.

"Years before," he said, "our limit had been 250 millirems a week. But when we began remodeling the B reprocessing plant they wanted to make it 500 and we'd have had to absorb it, if it was convenient for them, all in one day. Working in a hot zone is like a doctor using sterile techniques—if you do things right there's very little hazard, but they have to be done right. So we said 'No, the limit should be 200.' We finally settled on 300 maximum in any seven-day period, 2,000 in any quarter, and 3,000 in any year—compared to the 5,000 AEC limit. We made them give burned-out people other jobs outside the hot zones, or else pay them the rest of the week in any case. We got those limits without even striking, because our people simply quit each time they reached 300, and when the managers called it insubordination we said, 'No it's not, we're just working by *our* rules.' "

Alden came here in 1943 from Louisiana, left to go into the Seabees, then returned again in 1948 for an eighteen-month job that ended up keeping him here ever since, splitting his time between on-site work and union business (he received journeyman's scale in either case). Though he never liked traveling, he had little choice when construction slowed in the 1950s, and after working for a while in town as a regular plumber, he left in 1957 to do pipe fitting in a Fontana, California, steel mill. When his daughter came down to visit and wanted to go to Disneyland he couldn't break from his twelve-hour-a-day, seven-day-a-week schedule to take her, but it was also the first time he ever made $500 a week. "I felt so rich," Alden said, "that I played banker and loaned money to all my friends, the construction travelers. I got it all back too, $5 here and $10 there over the next few years, and they'd drop it off at the house with notes saying I'd know who it was."

Alden looked around his living room with its comfortable chairs, fireplace, and the shelves lined with book club editions of Daniel Defoe, Mark Twain, Aristotle, Plato, Gustave Flaubert and Robert Frost. (He had spent a lot of his spare time reading them and until recently had the *New York Times* mailed to his door.) "We all live too high," he said. "Who ever heard of a plumber having a layout like this, plus two cars and a boat. Look at the apprentices' parking lot . . . All those macho pickups, four-wheel drives, swept-wing whatevers, and they love them more than anything in the world. When my son was growing up, the union was so tight I couldn't get him in even though I was the local president. Now there are 550 apprentices for under 3,000 journeymen, and instead of running three to four months a year like normal construction, this project goes on and on without ending."

Emphasizing that the WPPSS problems were caused by poor management, not by the men who built the plants, Alden recalled when, in 1964, Kaiser Engineers exceeded what they'd bid for work on N's Hanford Generating Plant. WPPSS refused to pay, defended themselves from the beginnings of a Kaiser court suit and settled for a fraction of the overrun cost. He told me about the improper scheduling and the workers standing idle. He defended the QA and QC inspectors being in the same union as the workers whose welds they checked by saying, "It's better rapport. No one has reason to put out faulty workmanship, and if you go anywhere in the world people know Local 598 for the quality of what we do."

Though Alden wasn't sure why people became antinuclear, he suspected many of the opponents were "theoretical purists," and repeated the standard phrases about "Jane Fonda and Robert Redford backpack types" who talked about the horror of atomic by-products but never mentioned uses like those of the isotopes with which they had done a scan when he had meningitis. Three Mile Island, though it did disrupt WPPSS bond sales, didn't worry him "because even with Murphy's Law working all day the plant still didn't blow." And though he admitted not knowing much about waste disposal, "as soon as someone gets smart enough to use the stuff we'll have a gold mine."

Alden did, more than anyone here, make me aware of the constant pressure to use workers as atomic fodder, explaining, "If you can double the radiation limits, you halve the cost by using fewer people." Given situations in which workers could receive a week's dose in fifteen to twenty minutes, the more plants the industry built the greater reserve force would be needed even for routine maintenance (it took between three hundred and five hundred pipe fitters, for instance, just to handle N reactor's regular summer shutdown). "The real Pandora's box," Alden said, "is the 200 Area with all its separations and processing complexes."

But when he compared costs for electricity generated from alternative sources with the 3½¢ per kilowatt price tag for atomic power, his 3½¢ figure used WPPSS projections made before their construction actually began and budgets rose more than fivefold. That radioactive isotopes would not percolate from the 200 Area to the water table in less than the thousand years he cited was highly arguable. And although the fitters honored the strikes of other unions, it seemed a strange solidarity that while Hanford companies generally kept their workers below the same radiation limits of 300 millirems per week, 2,000 per quarter and 3,000 per year, the weekly dose could be exceeded, with the permission of the individual involved, for all Area workers except members of Local 598. When I asked Alden about this, he answered only, "I wouldn't be so arrogant as to say we have the power and the others don't . . . but we just saw the opportunity and took it."

Alden said he considered the HCA lockout little different from any of the other fights he'd been part of in forty years as a union craftsman. But among the young workers it seemed to presage, if not a death, at least a curtailment of happy-go-lucky boom times. Since the dispute ran five months, many ran out of money, slid into debt, and began selling first their fine toys and later near-essentials. Although as Alden said, "Even Ferguson admitted labor got a bum rap in this dispute," the workmen saw a crackdown on them by WPPSS management, the NRC inspectors, the media, and of course by the host of nuclear critics. And they worried that their magic golden project might end.

Although mistrust of one's employer is a given of corporate industrial America, the five-month layoff clearly increased tension. When the young workers questioned who I was and what I wanted in talking with them it was with a fear and, at times, a hostility I hadn't witnessed before. When they discussed how WPPSS was intending to screw them, it was with a certainty the same speculations had never carried before. When one, in a drunken brag, slashed open his ID card to show a copper imbed he explained would "soon be used to keep track of us," it was an exhibit of paranoia which would, the year before, have been unthinkable.

Two days after Craig and I met at the Cosmo Angus (our visit had actually taken place right after the labor dispute ended), I visited him at his two-room Kennewick apartment. Though he lacked the usual giant TV, he filled the space with a water bed, brown beanbag chairs, a large stereo, a bike, his ski equipment and several cameras. He showed me another of his *Herald* letters, on the antiparaphernalia initiative, which suggested that if dope did make you "dull-witted and slow," it was therefore "perfect training for government work."

Craig moved here from Los Angeles thirteen years ago when his father, who now lives in Pasco, got a job first at N reactor and later in finance scheduling for Westinghouse's experimental fusion project. Craig had begun working summers in the Area four years ago at age nineteen (his first job was helping fitters by wrapping plastic around the hot zone valves they removed), and had used periodic six-month spells there to finance his way through Washington State University.

We left the apartment to see a carpenter named George, who had drawn unemployment and done off-the-books remodeling for a good part of the work stoppage, until he and his wife Susie had accepted her parents' invitation to move back in with them. We sat in the basement family room while George watched the Seattle Seahawks football game. The men discussed skiing, cars and different construction jobs they'd worked. Susie ate Sugar Wafers.

They switched their attention back to the game. When Craig asked George if he was glad to be getting back to work, his friend answered, "Not really," then turned to me and explained, echoing my conversations with Lou, "It's not like building a dam or a custom home. You don't get to actually watch something go up and get finished. Since I like sales, I've been thinking a lot, actually, about going into real estate. But I guess I'll just have to see what happens."

"You know, a lot of skilled craftsmen leave," said Craig. "Maybe because they don't like the WPPSS notion that you get the job done best by cramming as many people as possible into the smallest space you have."

"Or maybe they get tired," said George, "from having to spend all day going 'A little over there, a little over here . . . No, maybe you'd better make it higher' (he mimed the motions of a laid-back supervisor). By time they come home all they can say is 'Honey, I'm exhausted, can you get me some ice for my finger.' "

"We have some bad job-related injuries," said Craig, and told of a friend who got hemorrhoids from sitting four weeks straight on a stool doing fire watch.

I thought again about the gallows humor that had succeeded tinkering as a way of life, justifying all frustration and distancing all questions of consequence. Even the most cutting smart-ass comments seemed to reinforce the basic devil's deal: The most powerful crafts receive the prime trickle-down wealth, the rest take the leavings and all cede control over what they, the direct producers, manufacture. You can call the young WPPSS workers stoics: They belabor their frustration endlessly in private discussion, yet return each day to accept the admittedly absurd and irrational way things are, and therefore always have to be. You can call them pragmatic: They brush aside ideals and utopian hopes as if they were so much smoke from the dope they constantly consume. You can call them realistic; but I wondered if their realism couldn't have made a private peace with almost any situation. When I mentioned Sam Volpentest's dream of twenty-five reactors lined up pretty as pearls along the Columbia, Craig laughed, and said, "They'll probably all be little WPPSS II's, but why not put twenty-five or even forty-five? The Tri-Cities deserve them anyway."

8

Revenge of the Pointyheads

When I boarded a 7:00 A.M. bus to N reactor, the workers waiting to go out to their Area jobs evidenced no special aura as they dozed, scanned daily paperwork and read westerns, mysteries and the morning news. I entered the gray industrial building past the plant security station and a sign warning THEY CAN SURPRISE YOU—WATCH OUT FOR SPIDERS AND SNAKES. In the control room, I watched the reactor "go critical" (begin a self-sustaining atomic reaction) after a summer halt for repairs. High-pitched tones pinged steadily as damper rods were pulled. Operators watched gauges, adjusted dials and stood ready by gun-handled backup manual controls. A supervisor wandered around, checking figures against his data tables. Lynn, my PR guide, explained the workers' concern about the reactor "scramming" (shutting itself down in a quick automatic halt) and forcing them to begin the seventy-two hour startup process once more, then compared the tension to that in a maternity ward. But the edge of excitement I'd expected was lacking in these casually dressed and casually postured men and women whom I watched; they seemed nearly as bored as the radiation monitors sitting down the hall playing cards beneath a girlie calendar.

I stood for perhaps a half hour, realizing how much of a reactor operator's task was simply patient waiting, then followed Lynn down the corridors to the office of Director of Reactor Operations, Don Lewis. The rooms we passed were decorated sparsely with charts, graphs, shelves full of technical manuals, and little else. The walls held dispensers with free Sight Savers. In one area, radiation monitors painstakingly checked the wheels of a cart and the overalls of a technician who'd just returned from a hot zone. One bathroom's anonymous scribes proclaimed, "Black is beautiful, tan is bland, but white is the color of the big boss man," and answered the question "Know any good whores?" by listing United Nuclear, Exxon, Rockwell and WPPSS. When Lynn explained that N plant was different from Three Mile Island or Oregon's Trojan plant in being "primarily a production reactor," I didn't think until much later that this "production" could be of plutonium for nuclear warheads.

Lewis decorated his office with a chart outlining the mock organizational structure of some immensely bureaucratic company, a single ivy plant and this motto—"When a decision is made *follow it* with a positive attitude and can-do spirit." He sat with a general's erect posture and explained, when I asked about Three Mile Island, that the radiation doses received in the plant's surrounding area weren't dangerous.

"But what threshold level is safe?" I asked.

"You'd have to talk with a radiation monitor about that," Lewis said, then cited the by now familiar comparisons with doses received on a transcontinental jet flight or in high-altitude Denver. "Progress means mistakes," he continued, "but man is continually getting to be a healthier animal. I can't prove that nuclear power is safe. It can't be proved it's harmful. But my basis for judgment is that I've been working a long time in the industry."

"And how do you feel?" interjected Lynn.

"Grrreat," said Lewis, just like Tony the Tiger in the old Frosted Flakes commercials, and the empiricist's conclusion—"If I'm not dead this must be okay"—made irrelevant, as always, any further talk on the subject.

We discussed war and nuclear weapons in general (it was, ironically, the twenty-fourth anniversary of Hiroshima) and Lewis went from sounding almost like a pacifist ("If we're not going to use atomic weapons why make them? It completely terrifies me that our regard for life is such that men can even think of something like the neutron bomb") to a stalwart realpolitik believer ("Since we have enemies, we have to leave judgments to military leaders, not weaken ourselves so we get run over by a bunch of ruffians. If government experts who know more than I do say we need a neutron bomb, I have to back them up"). He ended the discussion shaken by the implication he might in any way be a dissenter from national policy, easing only after Lynn switched the subject by asking me what I thought of Ralph Nader. I said that I respected his integrity, though his style was a little too monklike for my taste, to which Lynn commented, "We wish he were a Trappist, because they take vows of silence."

Outside, pipes leading to turbine generators shone silver overhead, evoking the blur of night streamliners. Cooling water (from a nonradioactive loop) shot roaring geysers as it was vented before being emptied into the Columbia. A three-story golfball-shaped structure held liquid wastes before their initial processing. But all this could have been any industrial complex; the sole jarring presence was the incessant buzzing from an adjacent electrical substation where high-tension towers, resembling giant Hopi kachina dolls, brought electricity in from BPA-linked facilities that included Grand Coulee Dam, and sent it out from N plant to a power pool that served Oregon, Washington, Idaho, Montana and California.

As we waited for the bus to the 200 Area, a security guard explained to me that although the local deer came to be fed (the Hanford Reservation also serves as a federal wildlife refuge), they tried to keep them away "because you know deer don't know how to read and they might ignore a sign and eat something contaminated." Four joggers headed out on a lunch-hour run across the desert. The bus arrived, filled with a dozen employees, and headed off past the Gable Mountain site where Hanford engineers were drilling test holes for a possible national radioactive waste repository.

PUREX (Plutonium/Uranium Extraction) is an 1100-foot-long canyonlike building where spent fuel elements are moved by a shielded crane through cells containing chemical solvents, then broken down into the radioactive precipitates that include the most dangerous long-term "wastes." The building was on standby when I visited, as it had been since the old reactors shut down, but it was due to be remodeled and started up once again as a major plutonium reprocessing center. I stood there in a cramped repair space, while solution-bearing pipes—which ran through a massive concrete wall to and from the tanks containing the solvents—clanged and echoed as if they were those of some gigantic tenement steam system.

Behind the three-foot leaded-glass windows of the B reprocessing plant, robot manipulators welded highly potent cesium extracts into metal storage capsules. With their elaborate counterbalances, the manipulators were elegant, futuristic pieces of machinery, yet on a wheeled cart in one of the hot cells rested a ceramic jug hardly different from ones manufactured three thousand years ago. I put on protective yellow galoshes to enter the area around the thirteen-foot-deep cooling and shielding pool where the capsules containing the cesium were stored—then, with the lights off, witnessed a pale blue-violet glow resembling the color within a Bird of Paradise blossom. Called the Cerenkov effect, the color was created in this case by beta particles moving through water. Watching it is as close as one can come to actually seeing radioactive emissions. Staring at the diffusing lights, feeling I could understand the dreams of controlling the energy "brighter than a thousand suns," I thought about the hopes that intertwined with wartime fear when the atom was split.

But the significance of Hanford's technology becomes trivialized when images from it are used for businesses with such names as Atomic Foods, Atomic Body Shop, Atomic Lanes (a Richland bowling center), Atomic Plumbing, Atomic Health Center and Atomic TV Service; when nuclear symbols decorate banks and delivery trucks and the ads of a collection agency which brags "we don't use atomic bombs but our blast is equally effective"; when you can live on Proton, Argon or Nuclear

Lane, send kids to a school where the principal calls their mushroom cloud emblem "a symbol of peace" and praise God for the FFTF. The message becomes that of a bumper sticker which warned that a pickup carried radioactive material and stated "For safety's sake, don't follow too closely"—with the phrase curving up at the end, dancing happily on the paper strip as if it were a jingle on a TV screen, so that any fear it mandated was erased by the overwhelming boosterism: There are no better neighbors than neutrons.

In addition to the metaphorical language of machine tools, Hanford workers speak a new literal tongue that includes such slang terms as *Scram*, *crapped-up*, *Christmas trees*, *shapes*, and *hot zone*; acronyms such as the WOG T, em PHI and BPA SOOK V initial sequences decorating operation panels in the new WNP 2 control room; technical jargon describing "power excursions" and "criticalities"; and the bureaucratic double-speak of "outages" and "incidents" and "process upsets." The phrases could—since they describe processes which didn't even exist thirty-five years ago—signify this work as special and worthy of at least high respect, if not fear. But they're repeated too often. They refer largely to minute components of a system too vast and complex for easy comprehension. They end up simply reinforcing routinization.

The roots of what one could call atomic banality go back to the reservation's earliest days. The old hands were certain their special mission was helping to save civilization from Nazi barbarism, and they used all their skill, stamina and inventiveness to meet the project goals. But given that most Hanford workers had no notion of what they were producing and given that their explanations of specifically nuclear dangers were confined essentially to vague comparisons with X rays, the hazards that seemed most salient were those depicted in the standard industrial booklets which showed cartoon figures admonishing "Play safe, play sane" and getting mashed between walls and forklifts, sucked into flatbed presses for failing to pull down the safety guards or crowned by falling beams after leaving their hard hats on a siding. They were the risks involved in any mechanical effort.

Grand justifications dissipated further as Hanford's proclaimed purpose shifted from guaranteeing America's survival as a nation to providing electricity to run toasters and to manufacture aluminum for lawn chairs. And with the breakup of a consensus that at one point had even Pete Seeger singing the praises of the peaceful atom, the nuclear industry entered a beleaguered state in which both its official ideologues and most regular workers saw no choice but to insist that it involved no special risks, no unusual implications for human futures and no responsibilities beyond those held by the creators and operators of any technological system. The old hands passed these attitudes on both in everyday conversations and formal training programs. And the more they took special precautions, filled out triplicate paperwork and followed elaborate checks, the more they saw these acts as rituals guaranteeing that what they were doing was both safe and unexceptional.

But because they viewed atomic power as nothing extraordinary, the old hands mistrusted and resented the web of controls and procedures that the industry developed both to deal with immediate safety problems and to deflect opposition criticism.

"If they regulated cars like they do reactors," explained Bob, a Hanford tinkerer I'd met at the Kiwanis luncheon, "you'd need dual puncture proof tires on all wheels, a steel engine enclosure, twin radiators with two separate water pumps, at least eight instrument and control systems and an emergency diesel-generated power system with automatic switches. The car would run at 10 miles an hour getting 2 miles per gallon of gas. You'd have to stop if a warning light blinked, pull over and spend half a day writing reports."

I'm condensing his list, for, expert engineer that he was, he sketched out two dozen more features that just began to approximate the bureaucratic load placed on the nuclear industry—a load that has made those who built this place feel like so many Gullivers bound and assailed by the Lilliputian minions of the regulatory agencies. There were, of course, obstructionists in the past: Bob told of the time he bought a used remote control forklift for a fraction of its value from U.S. Rubber, equipped it with special wrench systems so it could be used for cleanup in contaminated zones, then was almost fired by a supervisor, "who mistrusted all decisions he didn't personally make," for acting without sufficient authorization. But that was a single, petty man whose attitude soon got him transferred out. "The bureaucrats now," Bob said, "think they can run everything from the seventh floor of the Federal Building. They've played divide and conquer ever since GE pulled out and left us with a half dozen contractors who the AEC, or whatever the heck it's being called these days, runs all over with talk about who's complying and who isn't. Being treated like little kids really takes the spirit out of all of us who built this place. But the higher-ups lower the boom every time the media screams. Because they don't want to give the newspapers anything to talk about, you end up double-checking twice as much and spending half your time just answering questions. If we get anything at all done these days, we're lucky."

Bob had built his own hovercraft—which was inspired by an *Argosy* magazine photo and "by a desire, after seeing Sputnik, to become a bit airborne myself"—using parts from crashed drone aircraft that he'd bought from a nearby military firing range for the 10¢ a pound scrap aluminum price. Its mufflers were irrigation pipes stuffed one inside the other with fiberglass in between. If the machine wasn't perfect, it worked well enough to fly, and Bob kept testing and rebuilding to improve its design.

Bob knew, of course, that reactors needed far more stringent tests than his space age toy. But the media and environmentalists wouldn't allow the nuclear industry to slip even once along the way to achieving the immense benefits it could generate; they expected projects to run flawlessly without so much as burping or coughing.

"Du Pont and GE were both safety-oriented," Bob recalled. "If you screwed up once you could get fired real quick. Everybody knew it and we never had anyone injured on any jobs I managed. But no sooner did I get put in charge of monitoring the waste tanks [working, under the diversification contract, for Isochem successor Atlantic Richfield Corporation] than I spent half my time writing up occurrences instead of working to reduce the incidence of failure.

"I was in a meeting with my boss once while they were repairing the fire alarms. When they went off with a false signal, the firemen came in anyway, said we had to leave, and because we insisted on staying, wrote us up. So I circulated a joke report explaining the incident as 'Logical Behavior.' "

"An ARCO department manager," began the report, "behaved in a logical manner during business hours at the 2704 East building—200 East area. He failed to believe an emergency condition existed and failed to flee from the building. It was not until he recognized embarrassment on the faces of executive personnel present that the employee realized the danger was real . . ."

The document, written on a regulation form, continued with a subheading for "Temporary Corrective Action": "Remind employee to respond to alarms regardless of circumstances." Under "Permanent Corrective Action" it concluded: "At any time in the future that the employee appears to have gained some self-confidence and/or tends to behave in a logical manner, show a copy of this report to him and all of his associates."

There is disagreement on which of the safety rules are necessary and which constitute needless harassment. Lester Dumont told me about writing up a man working under him who'd knowingly violated guidelines on entering radiation zones: "And he's been my enemy ever since." Because it was hard to stay alert staring at reactor dials on the midnight shift, Lester felt "you need close supervision to keep reminding operators who get inured to routine. You need good Quality Control on every part entering the plants and inspectors to make sure the Quality Control is done right. You have to check each system that's critical—which doesn't mean the office air conditioning, but does include nearly everything else." Though nothing was 100 percent safe, he felt that if you did have tight enough controls, the nuclear industry could be a reasonable venture.

For most Hanford veterans, though, the regulatory bureaucracy is seen as malignant and overinflated; most rules are unnecessary cases of what Herbert Marcuse called, in a far different context, "surplus repression." Tinkerers' skill should be sufficient guarantee of safety.

"The cutting edge now is devoted to writing up EPA specs," said a former manager, "and that doesn't get us into the new postindustrial age. The people who are results-oriented get slowed. The environmental movement had some points, but they don't know when they've won a victory. They want to push to the last tenth of 1 percent, mutilate the corpse and stop productivity."

"When I turned sixty-five, they made me retire from the Westinghouse fusion research project," said a veteran here since the first Manhattan Project days. "So I worked the past five years through a job shop on a series of six-month contracts, and now the DOE says job shops are for temporary people and five years isn't temporary. I'm trying to go through Battelle, which has no age limit, and have them subcontract me back to Westinghouse—but why should some silly clerk with a holy expression on his face have the right to decide that, when I know more than I ever did and can ride a horse 20 miles across the desert, I'm no longer fit to work?"

"We used to arrange contracts verbally through people we knew," said Sam Beerman by way of explaining why he closed his own small placement service. "Then, in the name of being democratic, they created a maze of formal procedures which ended up making things almost impossible for everyone except companies large enough to hire regulations and negotiations specialists. Just like the EPA, it ended up driving a lot of small people under."

"They send young punk inspectors fresh out of school," said another Hanford veteran, "and they expect them to know from books how an atomic plant runs. But they don't, all they have to go on are paper rules, and they wave them in our faces as if they were collecting bills."

The old hands' frustration extends not only to the government regulators, but also to constant and unpredictable changes in management. Since GE pulled out, Hanford has seen an almost biblical succession of contractors, all with their different style, codes and organizational approaches. Taking the 200 Area chemical processing and waste management facilities as an initial example, Isochem—that odd combination of airplane manufacturer Martin Marietta and tire company Uniroyal—arrived in January 1966, only to turn the contract over a year and a half later to the ARCO oil company. The N reactor had been operated jointly by Douglas Aircraft (later McDonnell Douglas) and United Nuclear Corporation. But in 1973 Douglas pulled out, taking with them a research facility, Donald W. Douglas Laboratories, which they'd brought in as part of an attempted economic diversification effort. In 1975 Boeing Computer Services replaced Computer Science Corporation for handling the Area's data processing. In 1971 ARCO took over not only the 200 Area, but also the site support services run formerly by ITT—including the Hanford Science Center, the buses that took the workers to their jobs each morning and the atomic laundry that washed contamination off the protective coveralls. In 1977 ARCO was succeeded in all its projects by Rockwell Corporation.

Immediate tasks and immediate supervisors remained constant through most of these changes, but the shuffle of corporate titles, corporate stationery and corporate rules helped create a general feeling that the single-track mission was no more. In the war years, as one more old hand explained, Hanford was a suburb of Washington D.C. and of Du Pont's Wilmington, Delaware, headquarters. "Under GE, we were a suburb of Schenectady, New York. We've never been a city and never our own town, but now—with all the different contractors and regulating agencies—we don't know where to look to understand how we're supposed to operate."

Present-day Hanford confounds not just old-timers frustrated at no longer being able to build without check, but also men such as Western Sintering's John Rector, who have created their own niches on the reservation's fringes. "Area wage scales are so high," Rector explained during our initial visit, "that I end up spending $30,000 monthly to hire the exact same number of man-hours for which my St. Louis competitor would pay $20,000. I take employees in, teach them to be skilled mechanics, then their names come up on $14- or $18-an-hour WPPSS lists and they call some Saturday afternoon saying 'Johnny, I'm awful sorry, but they need me on Monday and I just can't turn down the extra money.' I tell them I'd do the same in their shoes, which I would, but it means we can only hold on to a few top people. Since Hanford is less than 1 percent of my business, I'd relocate in a minute if it weren't for the friends I have here and my family being settled and comfortable."

Rector didn't want to end up like other edge-of-the-Area companies that folded or moved elsewhere in part because the hyperpaced growth of the government-subsidized economy had jacked the wage scale so high. He mentioned the Azurdata computer firm which once occupied much of a mini-industrial park by the Richland airport; HUICO, a Pasco piping manufacturer "who saved $4 per man-hour by moving to Salt Lake City"; and Automata, which had been making automatic grading systems for classrooms until it relocated to Puget Sound. He dealt with the problem in his own business by designing new automatic feeders and loaders that enabled him actually to raise productivity while dropping the number of employees from twenty-eight to fourteen.

When I drove out to see Rector at the Tri-Cities Raceway, it was one of those bright, clear farm afternoons which people had longed for nostalgically when they'd voted in Ronald Reagan as president just a week before. Going first through the town of West Richland, I drove over an old railroad bridge that displayed a sign saying "Riders, please dismount and lead horses across," then passed a cluster of thirty-year-old houses and taverns, the white stucco building housing the Atomic Radiator garage and a new orange mini-mall. I continued several miles farther into open desert, and after entering the raceway area, spotted Rector in his jeans, sweatshirt and blue knit cap, nailing siding on a shed which housed pumping equipment for an adjacent recreational vehicle park he was constructing.

"I wanted a fun course for the drivers and spectators," explained Rector, showing me the track—with eight-foot-high banks and three curves instead of two—which he'd designed himself "from when I used to race motorcycles back in Illinois." But although the track turned out to be the fastest half-mile course of its class in the United States, and although it held more people than any in the region except the one in Spokane, it had ended up losing money. The reason, Rector explained, was that "the same people who someplace else would spend $10 to watch an auto race, drop $200 instead and go to the Coast, to Las Vegas, or to the mountains. Look at all the campers, boats and RV's here. You couldn't take off like they do and neither could I. But to these people it means nothing. Drivers love the place, but they end up racing to half-empty stands."

With Tri-Cities land prices booming, Rector still did well on his investment and ended up leasing out track operations to an outside contractor. This freed him to take pride in his technical accomplishment, and he described how the track's unusual configuration was partly because of the terrain "and mostly because I wanted curves of three different sizes to make things a little interesting. If a track is too easy, the guy with the best equipment wins automatically. I like races best when half the cars could possibly take them, when the leads keep changing and when they're a challenge."

As always, he conjured up the lone man against the elements: the type of entrepreneur revered by capitalist legend for mixing his own sweat and muscle power together with machine oil, steel girders and, of course, the labor of loyal workers—then building his own empire strut by strut. Rector locked out a machinist union attempt to organize Western Sintering, not, he claimed, because he would have had to pay more money, "but because they wanted to control me with a closed shop and tie me up in endless paperwork. Life's too short for that." He believed any man with gumption would work, strive and succeed. He cared more for what he could build and call his own than for the comforts wealth would provide.

Rector finished putting up the shed, explaining how he'd made the last section removable so he'd have easy access in case the pumps developed problems. We drove to get a beer at West Richland's Coney Island Tavern. I asked Rector what he'd do after finishing the pumphouse.

"Probably redesign the automatic loaders," he answered.

"And after that?" I asked.

"And after that," he looked amused, "there's more ideas than I'll ever have time to do. I'd like to get into windmills. Even the guys doing the big Boeing ones have the wrong designs, the wrong installations and the wrong locations. I went down the Columbia Gorge [where the river runs through steep cliffs between Tri-Cities and Portland], took pictures and anemometer readings and found a place their figures say is worthless but I know is perfect. My design's not patented yet, so I don't want it coming out in print. But their million-dollar, hundred-and-fifty-foot long propellers aren't the answer. And if you do it right, you can end up generating more than the dams."

Windmills aside, Rector had no problem with Hanford or nuclear technology in general. He complained about "the TVA dam blocked by that snail, whatever it was," and said WPPSS was nearly stopped because "some ecologist" claimed the plants might affect the local gopher population.

"You've probably seen figures about industrial production dropping," he told me. "They don't say that the reason's because for every five men working there are three more just looking over their shoulders."

Bringing the subject back to Rector's own projects, I suggested his wind idea might be able to get Department of Energy funding through Battelle's renewable energy program.

"Wouldn't want it," he said, shaking his head. "I know Kirk [project director Kirk Drumheller], and he's a good man. But I wouldn't touch that paperwork. I don't mind paper if it's to draw blueprints on like I did designing the raceway or the sintering machines. But life's too damn short to spend it writing endless studies to go in someone's file drawer."

Hanford's old hands believed the wave of restrictions and regulations, which came in the wake of their environmental opponents, was based largely on ignorance and resentment. They resisted by throwing the young inspectors out of their offices or by writing required logs and reports on disarrayed scraps of paper. They griped about being constantly hamstrung just as the young WPPSS workers did about logging endless hours on projects that might never be completed. Although the pragmatic tinkerers did often judge their surrounding environment instrumentally—valuing it only for the "resources" that could aid their mission of building—the government bureaucrats in turn seemed to value the tinkerers only for their compliance with petty rules.

At one time faith in the urgency of the task justified enduring makeshift accommodations, constant scrutiny from Military Intelligence and ignorance about the ultimate product. Now the men hoped that what they began in the early glory days would be allowed to continue, and that the language of invention they had spent nearly thirty-five years learning would not be capriciously banned. Although physicists elsewhere did the pioneering theoretical work, Hanford's old hands had once believed that their own creations and their own day-to-day labors would be part of bringing about an era in which the world could be shaped and altered as never before. Now recollections of designing separator cells, centrifuges or reactor claddings served largely to buttress a besieged faith that their efforts had indeed been noble and worthwhile.

PART III

OUR NUCLEAR FUTURE?

"The unleashed power of the atom has changed everything except our ways of thinking. Thus we are drifting toward a catastrophe beyond comparison. We shall require a substantially new manner of thinking if mankind is to survive."—Albert Einstein

9

A Job Like Any Other

When United Nuclear representative Gladys Lowry talks with fourth and fifth graders, she explains, "You know when you come home, kids, you're all dirty and your mother says you have to take off your shoes and sweatshirt—well, nuclear safety is the same sort of thing." Since school was out of session during my initial visit, she gave me a surrogate presentation in a basement locker room of the N reactor administrative complex. Gladys began by putting on a protective suit that included a mask, a cowl, three layers of cotton coveralls and the white canvas shoe covers which she tells the kids are "Santa Claus boots." With the help of Lynn, my PR guide, she recounted the standard lecture. I watched, half critical journalist, half wide-eyed ten-year-old.

Gladys had worked here fifteen years as a dosimetry specialist, and she had talked on behalf of the industry to over 10,000 people in the past year. Because the coveralls were uncomfortable over slacks, she wore shorts and a tank top, as did many of the women who suited up in hot zones. The contamination suit fastened with Velcro, "because you don't want to be hauling and tugging at zippers." Masking tape wrapped around the edge of each overlapping garment, forming a tight seal and leaving a folded-under end tab to ease quick removal.

Gladys put on a rubber cap and—after blowing them up first to check for pinhole leaks—a pair of rubber surgeon's gloves. She placed the cowl over her head, which made her resemble a sister in some futuristic Order of St. Fermi, added the yellow breather-equipped Mine Safety Appliance mask, then handed it to me so I could look out with cramped underwater vision.

She described how workers around radioactive liquids wore heavy-duty "canary suits" resembling booted and wader-equipped versions of grade-school yellow rain slickers.

When I asked Gladys whether she'd heard about contaminated waste water running into the ground from where Rockwell launders the hot zone coveralls, she answered in an official, perfunctory tone, "I'm sure it's been taken care of." In terms of the then-recent Three Mile Island accident, she responded only, "I think enough's been said." And when I wanted to know what radiation level she, as a dosimetry specialist, considered dangerous, Gladys said she only dealt with measuring accurately the doses workers received.

She'd mentioned earlier her husband's death from leukemia (he'd been a radiation monitor), then changed the subject before I could ask whether his work here might have caused the disease, and began spinning her vision of how women needed unlimited energy as a prerequisite for liberation. But Gladys no longer seemed to be thinking of the bright nuclear future. She'd stopped talking about her upcoming trip to China, the way the desert wind sounded as she drove her Z-28 Camaro out to the reactor each morning, and about how—in phrases which sounded like those of the moral bear family in *Highlights for Children* magazine—safety here was like reading the instructions for toys, dumping muddy clothes in the hamper and not leaving skates where people can trip on them. She was blinking instead, the way a child does when they don't understand, and while I asked again how certain one could be that the doses received here were safe, she fingered a loop of contamination suit tape as if it were a rosary bead.

Given Gladys's sleek professionalism and nuclear surety, this momentary break in her composure surprised me. I thought again of her husband and of a study, initiated by University of Pittsburgh public health professor Thomas Mancuso, that correlated radiation doses received here with instances of a number of cancers including leukemia. If Gladys had been the wife of a worker dead in some construction fall or chemical explosion at least the tragedy's immediate cause would have been clear. But while her husband's death might well have had no connection with his Hanford work, and while she would insist unequivocally that it didn't, could she be certain her reassurances weren't presaging a similar future for others?

Though Gladys relaxed a bit, her expression remained stony. She removed the contamination suit article by article and explained tersely how the industry was so safety-conscious they even burned the sealant tape in a furnace with filtered stacks.

I saw Gladys again a couple of months later addressing a coffee klatch in a town 60 miles north of Seattle, where a referendum the following week would prohibit the building of a long-planned atomic plant on the site and make likely its probable relocation to Hanford. Sponsored by a nuclear industry-backed women's organization, Gladys suggested conserving energy by recycling egg cartons and turning them into planters, then warned that while solar and other alternatives were nice, in energy-short times we should not forget our friend the atom.

There are different levels of Hanford risk-taking. Being bombarded with gamma rays while spending your dosimeter-measured half-hour in a hot zone is different from handling radioactive materials with tools and procedures that supposedly protect you from all exposure. The first situation demands faith both in the harmlessness of doses beneath the allowable threshold and in the stability of official threshold limits which have dropped from almost one REM a week in 1934 to fifteen a year in 1950 and five a year since 1957; the second requires only that containers, glove boxes and general mechanical systems will maintain their integrity. If one works at WPPSS, several miles away from any radioactive materials, there is only an abstracted responsibility for plants which often appear as if they'll never get built anyway.

One could make the harsh judgment that Hanford's workers are not victims but executioners poisoning the environment and future generations. But the men, women and children here would be the first affected by any hazards of their enterprise. No one wants anything but safety. Why expect atomic employees to take more responsibility for the consequences of what they produce than do workers in any other industry?

To the men and women who strolled through Columbia Center, wearing their Area badges as casually as if they were decorative pendants, the atomic processes required perhaps special rules, but certainly no grand assessment of planetary impact. Some even made the risk a game, like the hot zone workers who wanted free time and burned themselves out prematurely by leaving their dosimeter adjacent to dental X-ray machines or other sources that would increase the amounts of radiation they registered. (Others were rumored to tamper the other way so as to gain more working hours, but as a Battelle dosimetry specialist said, "That's possible, but it would take a lot of technical sophistication.")

At times the risks were treated as a macho challenge, as in the case of a pipe fitter up from Texas to do some work for Battelle. Scheduled to take out some pipes in a hot zone, he'd been warned that the protective gear he'd have to wear would restrict his movement and be hot and uncomfortable. When he went in, instead of doing the job as quickly as possible and taking care to avoid any contact that could contaminate him, he grinned at his partner and did a chin-up on one of the pipes. Just to prove nothing fazed him, he did another and followed that with several more.

The radiation monitors were furious, of course, and made this completely evident when the pipe fitter returned outside to change. But as he was taking off his coveralls he dropped an earring on the floor, and before they could react quickly enough to check it, he picked it up, popped it in his mouth, said "Got to keep it clean" and grinned even more broadly than before. The monitors surveyed his mouth, his clothes, his entire body. Fortunately, his contamination was minor and washed right off.

Perhaps the distance between Arthur Compton's coded message announcing Fermi's initial chain reaction success—"The Italian navigator has landed in the New World"—and the attitude of the Texas pipe fitter or the commands of green tags—which proclaimed, even on reactor building toilets, "Custody of WPPSS. Start up only under the direction of WPPSS start-up personnel"—was that between a laboratory and a production factory. When the Z plant training manual suggests that workers view their operation "as either a complicated scrapyard where we clean up trash to collect plutonium . . . or a plant that refines high-quality chemical products," it presents an inducement highly different from either the allure of the Cerenkov glow or the imperative to save America from the German war machine.

A dulling of context also develops from the specific tasks performed. Hanford is an environment in which security requirements have traditionally mandated that only the most task-specific aspects of the job will be discussed, and in which nothing in the fragmented component systems can give a sense of the product's ultimate implications. The old hands are isolated from people who work for any institution except the all-paternal Area. Their machine language is so safe and familiar they begin to shy away from any other languages, from any other approaches. Given all this, it should have been no surprise that many here responded to questions I asked about the nuclear cycle by stating that they knew I was wrong, but my apprehensions could better be refuted by people in other departments. Or that one young woman I met knew that her husband worked "at the Area" but had no idea for which company or on which project. Or that a former head of Battelle was unaware of his own research lab studies indicating higher-than-background radiation levels in the Columbia. Senses grew so narrowly focused here, and the system they had to comprehend was so immense, that whether you bled and prayed for the industry or merely punched in every day, it became almost impossible to assess the import of what you did. It was easier to retreat to inventing for invention's sake or to logging your time for a paycheck.

Leafing through the Z plant training manual, I thought about its real lessons. True, I now knew how far apart different amounts of plutonium should be placed, how the oxides were separated out, and even why—because of the potential for theft by "hostile factions"—security and safeguard requirements were the most stringent of any project Rockwell had here. I recalled a vague caution or two regarding personal contamination. As with any infinitely detailed rulebook, I learned primarily that if I violated some minor regulation I could lose my job. No doubt the trail of paper double-checks did facilitate industry safety. And if the directions could anticipate every contingency, simple obedience would eliminate human error. But for every situation in which redundant procedures made an evident difference, there were fifty others in which they appeared so blatantly absurd they caused worker resentment. Formal rules put the fear of punishment first—which is why, when a WPPSS pipe fitter saw me wearing a white hard hat and writing in a notebook, he took me for a visiting inspector and stopped his work to keep an eye on exactly where I was and what I was looking at. Formal rules leave little room for responses outside prescribed channels—and left the Three Mile Island operators trapped between conflicting commandments never to turn off the emergency core cooling system and always to prevent pressure buildups that might rupture the main reactor piping. Formal rules separate what workers are required to know from what is considered simply not their affair—so the "just a job" attitude grows more and more prevalent at Hanford.

The backlash to proliferating control is resentment. The Hanford Contractors' Association attempts to discipline an unruly WPPSS labor force, and graffiti goes up—"Fuck the HCA," followed by the complementary elaboration "Why not?

They fucked us." The WNP 2 pressure dome becomes decorated by a flip drawing of the R. Crumb cartoon character Mr. Natural giving his countervailing "Lesson in Life." The projects themselves, say the workers, still "won't be completed until the height of the safety forms equals the height of the reactor containment buildings."

The old hands resent the bureaucratic regulators who, in their view, have collaborated with environmentalists and other dissidents to bring a once proud lumberjacking, iron-slinging and empire-building nation to its knees. In reaction to this they have stressed that their work presents no exceptional hazards and at times have become as careless as the twenty-five-year veteran chemist who decided he'd transplant plutonium to a San Diego lab by shipping it inside an ordinary suitcase on the commercial airline flight he was taking. His intentions, despite his massive breach of security and safety procedures, were the honorable ones of getting the job done in the quickest and most efficient fashion.

Hanford's younger workers, though, carry no such allegiances. While nuclear critics worry about the stability of the political and economic entities necessary to safeguard radioactive materials for periods of up to a hundred times as long as recorded human history, the institutions of daily culture have changed in a manner which increases atomic risks even in the thirty-five years since the Area was begun. For many here who came of age in the period from the late 1960s through the 1970s, Hanford is just a giant institution no more deserving of loyalty than any other they disdained, perhaps even rebelled against, and then acquiesced to serve while still withholding half their souls. If sexual hustling is more important now at Z plant than when the facility was built thirty years ago, this is in part because the work now means far less to those who do it. When people get high going to and from their jobs each day, steal more tools than they can ever use or joke about how they accidentally destroyed $10,000 worth of this or that type of sophisticated equipment, it's because it now seems ludicrous to consider state or corporate projects as noble causes. Instead the workers play the game because, as pipe fitter Dave said, "You need money for gas, weed, rent and wine." And they prove they aren't company pawns by sooner or later breaking the rules.

The more Hanford jobs become sectored off and mechanized, the more they create an atmosphere of boredom. The more boring they become, the greater chance inattentiveness or sloppiness leads to mistakes. Sometimes the incidents are almost comic, as in the case of a security guard who found an abandoned wrench awaiting disposal in a hot zone. Without scanning it on any monitor, he tossed it over the perimeter fence, picked it up on the other side and placed it in his car. The wrench turned out to be contaminated. Monitors who surveyed the fence traced a radioactive trail to the parking lot. Since someone there had seen the guard, they asked where he'd gone after leaving, and discovered he'd crapped up both his own house and that of a girlfriend he'd stopped to visit before continuing home to his wife. When the incident led to each of the women discovering the other's existence, both ended up leaving him.

In a far more serious 1976 incident, operators at the 242 Z building (an adjacent section of Z plant) left a solution containing the highly radioactive element americium and the resin used to process it in a separation column for the duration of a five-month strike. Because the resin tended to break down and react in a volatile fashion with the nitric acid later added in processing, regulations prohibited the mixture being left in the columns for more than ninety days at a time. But the rules were ignored for column 14-A, and by the time sixty-five-year-old nuclear-process operator Harold MacClusky came in for his midnight shift after being back a few weeks, other workers had added the nitric acid. Within a couple of hours MacClusky noticed that the mixture in the column had an unusually dark, murky-orange color, and sent a sample to a lab for analysis. But before the results could come back, he heard a hissing sound and saw dense dark fumes rising inside the sealed glove box through which the column ran. He sent a co-worker out to call for help. He was looking in one of the lead-glass windows to determine what should be done when the solution exploded, blowing out the windows and glove ports and showering MacClusky with pieces of glass, resin, radioactive metal and nitric acid. MacClusky crawled from the room on his hands and knees, and in the process inhaled enough americium-laden fumes to ingest more radioactive material than any nuclear worker has ever done.

That MacClusky survived was due to months of intensive treatment in a special medical facility with lead-shielded baths, beds and operating tables; to massive injections of an experimental chelator that bonded to the americium and allowed him to excrete it from his body; and to the extended time period over which his radiation dose was absorbed (given his age, neither future cancer nor damage to reproductive chromosomes posed major worry). It took two and a half months before MacClusky's body had gotten rid of enough americium so people could stand next to him without special protection. Because of this and his freakish survival, he is now referred to as the Atomic Man.

At times, carelessness means lack of foresight. Z-9 was a crib—a trench with concrete walls and an earthen floor—which received contaminated waste liquid from the processing plants. Those who designed the cribs assumed soil particles below would trap and hold the dispersed radioactive elements contained within the solutions. Even the plutonium—of which, according to the government's conservative figures, there was between 25 and 70 kilograms (55–155 lbs.) by 1972—would be contained far enough from the water table to cause no risk over its 24,400-year half-life.

Unfortunately, as physicist Walter Patterson writes in his book *Nuclear Power*, "The procedure failed to take account of one factor. The soil beneath trench Z-9 did absorb plutonium as anticipated, but it did so selectively. Like the standard chemical separation technique of column chromatography, it separated radioisotopes into layers of different species at different depths. One layer not far below the surface was found to be dismayingly rich in plutonium: so rich that heavy rain soaking into the ground might, with its moderating effect, trigger a nuclear chain reaction." Accordingly, the Z-9 was dug up in 1972 at a cost of $7 million.

One could regard the problems of both MacClusky and Z-9 as stemming from a task-immersion which, in its inability to look beyond immediate needs and operations, complements the atomic industrial routineness. Difficulties could arise in any aspect of the atomic cycle from the mining of raw uranium through reactor operations, fuel reprocessing and eventual waste disposal and plant decommissioning. They could arise from improper planning, construction or maintenance of the complex systems that control the functioning of all individual components. They could reach crisis proportions when those systems fail and the workers monitoring them are unprepared.

Take reactors as an example. Thousands of people build each given plant. Thousands keep it running. They deal with thousands of minute parts, each of which can cripple a near-infinity of others to create what Daniel Ford, executive director of the Union of Concerned Scientists, has labeled "common-mode failures," for the manner in which single problems can affect a number of interconnected systems. For all that some workers focus intently on their tasks, others will inevitably be thinking more about the coming weekend's fishing trip, about the women (or men) they'd like to make it with, or about how much their head hurts from chemicals ingested the night before or even from some Agent Orange dose received ten years back in Vietnam. Because few special indicators distinguished components affecting the most critical systems from those plugging into office air conditioners, pretty soon all tasks become interchangeable. (This doesn't even count the off-site manufacture of parts, often by general industrial suppliers whose prime accounts have nothing to do with the nuclear enterprise.) And the more plants are built and go on line, the more the supply of skilled and careful repair craftsmen is strained by hot zone burnout.

A 1981 *New Yorker* article by Daniel Ford cited a dozen instances of emergency feedwater system failure that occurred in Babcock and Wilcox reactors before the Three Mile Island accident. Their causes ranged from a valve being accidentally stepped on by a maintenance person, to disconnect switches installed in narrow corridors being bumped into by passing workers, to a pump motor stopping because a loose nut had created an electrical failure. As always, the breakdown of major systems did not mean each of their parts simultaneously exploded or disintegrated; rather, single elements fused or snapped, froze, overheated or somehow otherwise malfunctioned—and then created their own chain reaction effects to disable the prime controls on the atomic process.

Amy and her Z plant colleagues were supposed to wear masks while opening the specially designed 55-gallon drums in which the actual plutonium containers were transported. But the masks were clumsy and no one used them (when Amy wanted to put hers on anyway for fear of being contacted, a veteran RM told her, "Look, if I don't wear a mask then you shouldn't either"), and it was just fortunate they encountered no contamination in the drums. Although neither Julie nor Robert caused the improper packaging that set off Z plant's plutonium fire, Robert made its effect worse by knocking the flaming can off the table with the fire extinguisher. A Battelle worker I knew made a mistake by picking up an unmarked container covered with a plastic bag—and, because someone else had improperly packaged it, spilled americium on his hands and clothing in an exposure that fortunately washed off entirely. And though the faulty Three Mile Island pilot-operated relief valve could not be blamed on the operators on shift, the choices made when a hundred alarms began flashing red and yellow lights and sounding warning buzzers were ones that furthered what came close to being America's most catastrophic nuclear accident.

As Boyd Norton, an AEC physicist for nine years, said in a 1980 *Audubon* article on Three Mile Island, reactor operation (and, as he explained later, most day-to-day nuclear work) is tedious and unchallenging until there's a crisis or emergency. "Then it gets too challenging in one hell of a hurry." After comparing the approximately $21,000-a-year maximum salaries of Duke Power's reactor operators with the $90,000 a year made by 747 jet captains, Norton explained: "Different industries apparently have different theories about compensating highly trained people for jobs that are responsible, potentially dangerous and essentially boring."

10

The Men Who Make the Bombs

There was a Hanford engineer whose first wife, a soft touch, convinced him to support a Calcutta eight-year-old through the Christian Children's Fund. When he remarried, his new wife considered the donation "a little sappy, but harmless," and thought nothing more of it until the child's regular progress reports stopped arriving.

"What happened?" she asked.

"I found out it was a blind kid," her husband answered, as if he'd been conned. "I mean, Christ, I was trying to help someone who could make something of himself, and would grow up to help his country develop. Instead they give me a blind kid! What's a blind kid going to do to solve anything?"

The engineer's harsh judgment didn't mean the Hanford men and women were any less generous than the residents of any other place, but his attitude was consistent with the instrumental approaches I found there.

Lewis Mumford, again in *Technics and Civilization*, has described military organization as "the ideal form toward which a purely mechanical system of industry must tend." It is a form, he says, which heralds "the quantification of life and the concentration on power as an end in itself." It denies all "knowledge of who and why, wherefore, for whom, and to what end." It inhibits all impulses except those useful in achieving victory.

In November of 1944, less than two months after B reactor went critical (and a year and a half after the first Hanford ground was broken), American troops entered the German city of Strasbourg, occupied the Physics Institute attached to the university and determined through the investigations of a special scientific team that the Germans were not even remotely close to developing an atomic weapon. There were a number of reasons for this, but one of the prime ones was passive resistance by a group of key German physicists that included Nobel Prize winners Werner Heisenberg and Max von Laue. Fearing that if they left their posts they would be replaced by enthusiastic servants of the Reich, these scientists stayed in Germany, but kept secret all research indicating that a nuclear bomb could be produced in time to aid the Nazi war effort, and they repeatedly denied that military applications were even worth pursuing.

When Allied atomic scientists learned that the weapon they were producing would not be a defensive counterbalance, but one we would use unilaterally, many questioned continuation of the project. They felt dropping the bomb on a civilian population would be inhumane and would set off a postwar global arms race. As seven prominent Manhattan Project scientists argued in a report to the Secretary of War (their study was called "The Franck Report," after the group's chairman, Nobel Prize winner James Franck), "the military advantages and the saving of American lives achieved by the sudden use of atomic bombs against Japan may be outweighed by the ensuing loss of confidence and by a wave of horror and repulsion sweeping over the rest of the world and perhaps even dividing public opinion at home."

But weapons created were seen by the military as weapons to be used. General Groves, the Manhattan Project head, pushed for even greater work speed as the war drew to a close. When Leo Szilard (who was also a major voice in the Franck group) made a final effort by drawing up a petition opposing the use of the bomb and gathering sixty-seven signatures of Manhattan Project participants, Groves declared the petition itself a "secret" document, and because such papers could only be transported under military guard and Groves said he could not spare any men to accompany it, he thus prevented its further circulation even among the top-clearance scientists.

At the same time that men such as Szilard, Franck and Einstein were trying to rein in the weapon they had helped create, they were unaware of the fact that our intelligence services had intercepted communications indicating that the Japanese were exploring the possibilities of surrender. And they were too innocent even to contemplate the possibility—suggested in later studies such as Gar Alperovitz's book, *Atomic Diplomacy*—that we might end up using the bomb as much to establish quick superiority in a forthcoming struggle with the Soviets as to put an end to the Pacific war. But aside from all the reasons nuclear weapons should not have been used, one must still ask why so little postwar debate took place on the decision to develop an atomic arsenal capable of ending civilization on earth as we know it. Instead of holding a public dialogue, the matter became a military, and therefore effectively technical, choice. Like the Manhattan Project, the AEC demanded an unquestioning obedience, a security establishment jealous of all outside its control, and a continued sweeping away of any cultures and communities—like those of the Pacific islanders who lived where the H-bombs were tested—that stood in the way of the all-essential mission.

Though without massive proliferation of breeders our planet's supply of uranium is far more limited than that of oil, the atomic industry still represents a refuge for those hoping to build endlessly without worrying about external limits. From the Area's beginnings, the promise of infinite energy—fueled by the reality of building weapons of unprecedented power—led the old hands to dream that their work might lay the base for an era which, as Clark Reitnauer put it, "would make all those utopian novels look like nothing."

Though the nuclear industry's massive scale and blend of military and corporate enterprise meshes easily with the operating modes of America's acquisitive and impersonal institutions, the old hands' strongest desires involved far more than the mere offering of service to the powerful. Although they accepted the realpolitik necessity for plutonium production, they envisioned the peaceful atom keeping America a fluid and free nation that would never need to choose between the red bogey of radical social change and the restrictiveness of a frozen class hierarchy. If their efforts pushed the technological frontier a bit farther, they could remain humble men, working happily with their machines and providing well for their families. Their labors would let us all keep our lives easy, comfortable and uncomplicated, while allowing distant leaders to carry the burden of making messy political and economic choices.

Hanford workers call themselves pragmatic realists. They say they deal with the world as it is rather than striving for utopian abstractions of how they might like it to be. They judge success in all ventures by what "works" and keeps the ventures going with maximum efficiency. Though they take pride, as they should, in their inventive abilities, they see little need to waste time evaluating the long-range implications of what they do.

Just as America is in general a culture in which most of us—only employees where we work and only transitory migrants in our communities—leave control of our world to others, the Hanford workers possess a range of vastly different sensibilities whose common thread is acquiescence to authority. The old hands who desire only to invent and to see their atomic mission bear fruit, have a far different attitude from that of the young cynics who log their hours, doing as little as possible, then come home to "real life," their own private time—and certainly a different outlook from those who, like the Germans termed "inner immigrants" by Hannah Arendt because they always privately abhorred the Nazis, mistrust the potential consequences of what they produce. Yet these inner immigrants go in each morning to participate halfheartedly, just as they do in other industrial enterprises across the country. And, taking Z plant as an example, when a "campaign" is called for and plutonium is reclaimed and shipped out for uses that include production and development of thermonuclear warheads, neither the old nor the young ever refuse their assignments on moral grounds. Even as Hanford's routine inures people to the possibility that the technology they work with might somehow run out of control and create some catastrophic accident, so the daily repetition and tedium remove responsibility for another possibility: that the systems might all work, that the switches and circuits might all connect, and that weapons whose plutonium was produced here might bring atomic holocaust to people the Hanford workers have never laid eyes on.

Can anyone claim specific responsibility for dropping the nuclear bombs? Should we say 350,000 persons died (combining the immediate toll with that of those who succumbed to delayed effects) because some maintenance person tuned a B-29 engine, because some factory operator put together the bulldozers that built the Hanford reactors, or because some farmer bought war bonds or raised hogs that ended up as pork chops on the Los Alamos mess hall table? But together with their Oak Ridge and Los Alamos colleagues, Hanford's old hands did have a far more intimate relationship than most Americans with the "devices" whose initial target became symbolic of an age. As late as 1965, Hanford had seven production reactors working solely to produce materials for nuclear warheads. Until the arrival of WPPSS, the FFTF and Battelle's research labs, all Hanford facilities except the dual-purpose N were exclusively weapons-related. From their wartime beginnings to the time when the last shut down in 1971 (with the exception of N plant), these reactors produced approximately 75 metric tons of weapons-grade plutonium* (165,000 pounds)—or enough for 7,500 Nagasaki-power bombs if used directly, and for infinitely greater destructiveness in their prime use as triggers for thermonuclear explosions. Almost 60% of America's plutonium warheads have been manufactured with Hanford material.

The Area ceased to produce weapons-grade plutonium during the 1970s, when its defense functions consisted largely of

*This figure, derived from waste disposal analyses made by Dr. Thomas Cochran of the National Resources Defense Council, could be up to 20 tons higher or lower.

managing wastes and keeping systems such as those of the PUREX reprocessing plant on standby. In the past few years, N reactor's spent fuel elements have begun to saturate the facilities available to store them, and preparations have started to get PUREX operating once more. Now, with the Haigs, Brzezinskis, and other born-again thinkers of the unthinkable insisting that our atomic arsenals are inadequate to overkill the Soviets, the warheads of MX, Trident, Pershing and Cruise will soon require massive amounts of plutonium for their trigger elements. In October 1980, the Carter administration issued a nuclear stockpile memorandum requiring a production increase, and preparation was begun shortly afterward to convert N reactor from fuels-grade to weapons-grade production (both can be used for bombs, but the latter is more "efficient") and to speed up the PUREX renovation.

When the first atomic bombs were dropped, Hanford's old hands were merely employees of a wartime facility, responding to nearly four years of terrible fighting and hoping the weapons they created would avert an invasion of the Japanese home islands. That America had to build and use them was part of the inevitable, perhaps tragic, progression of history. When the war years ended, Hanford workers knew they were producing an artificial metal, plutonium, that would be used in nuclear weapons, but they still did not discuss the implications of what they were doing. Defense Department officials explained the bombs were required for security. But this security fed, clothed, transported or housed no one, although those who worked for Hanford and other military or military-support institutions had to be fed, clothed, transported and housed by the rest of us. If, as the old hands hoped, Hanford's products were never used, the men and women here might still end up going in day after day, year after year, trying to meet an insatiable demand. Though most would die if the weapons were employed in even the most limited exchange, they approached their plutonium work—just as they did the atomic commercial cycle—as an industrial process little different from any other.

"Perhaps it would have been better if the bomb had turned out to be impossible," Clark Reitnauer said, when I asked whether he felt his work might end up laying the ground for a potential atomic war. "But if weapons like that are going to exist, it's better we have the bigger ones than smaller. Though I'd have preferred that neither Hiroshima nor Dresden was necessary, I can't say dying from radiation is worse than from a firestorm. I admit H-bombs do give me the chills. But you can't put the genie back in the bottle.

"I worry how this makes me sound, but I'm rather contemptuous of people who think human nature is nice. It isn't nice. It's savage. If you don't have civilization and fear of retribution people become animalistic." Reitnauer went on to tell me how his brother saw Detroit auto strikers "chase a scab into an alley and literally tear his arm off." He said that any group could become as bestial "with the proper provocation," and that he learned from growing up in mining towns "that if you don't defend your home no one else will do it for you."

"Right or wrong," explained Lester Dumont, "I never had feelings of contributing to any holocaust of the world. Initially, we all wanted to get the bomb finished before the Germans, and we did. Perhaps instead of dropping it on inhabited cities we could have given a demonstration for Japanese observers on some empty island. But producing plutonium after the war gave me no problem at all because I knew whoever, in this unfortunate world, had a greater ability to destroy would survive. I didn't feel the U.S. would be attacking people with these bombs. But we have to maintain a balance of power so other nations can't say 'Do what we want or we'll blow you off the map.' Would the Russians disarm just because we did?"

At times the old hands trivialized the potential effect of the weapons they helped create. "Atomic bombs are just bigger bombs than other ones," said Clark, and John Rector referred to the Eniwetok tests as "just another step in human progress."

"Are you going to return to bows and arrows?" Rector asked. "There was a time in history when they tried to outlaw guns. Nuclear weapons are just part of a natural evolution. Lasars have even greater possibilities. There'll probably be other weapons coming along we don't even know about."

Rector thought America was a large enough country that, if we dispersed our industry and population, even bombs devastating everything within a ten-mile radius couldn't knock us out. Atomic weapons didn't leave a person any more dead than bullets or firebombs. And as far as the possibility of planetary holocaust went, Rector assured me: "We'll also be developing countermeasures."

For the most part, specters of war retreated behind the immediacy of daily work and invention—behind dreams, like those Rector had, of how bombs could be used to blast new Suez canals; behind memories, like those Reitnauer related, of how "Hiroshima excited us because instead of someone else doing it to America, we were doing it to them"; or behind an understanding that, as Lester Dumont said, "If anyone had announced 'I don't approve of bombs and I'm not working on them,' all that would have done was cost them their position."

Perhaps the general attitude was summed up by Sam Beerman. "It was almost a peacetime industry," he explained. "We were proud to work for a major company like General Electric. We felt we were part of a well-run industrial enterprise with good management practices, good cost control and a good competitive feeling because the AEC would be comparing our cost and productivity figures with those of Savannah River [another plutonium production complex built in the early 1950s in South Carolina]. Some of us even went on recruiting trips, visiting different colleges along with other people from GE divisions around the country—and we explained plutonium as simply our product, just as light bulbs or turbines were someone else's.

"Remember," Sam emphasized, "our attitude was that the bombs were defensive, not offensive weapons. Building them wasn't warmongering. It wasn't like creating a machine for Hitler to use. It was doing a necessary job."

If Hanford workers treat nuclear weapons casually, they do so only in closer proximity to them than the rest of us who, whether or not we approve of their presence, manage to acquiesce readily to living alongside them. While working on this book, I attended a symposium on nuclear war sponsored by Physicians for Social Responsibility, which packed a 1200-seat hall at the University of Washington. After a morning watching slides of the devastated Nagasaki hospital and listening to speakers discuss in detail the effects of an atomic exchange—from shortages of antibiotics to death of most medical personnel and the destruction of a major portion of the earth's ozone layer—I stepped outside to see some journalist acquaintances. Though there was nothing wrong with any of our discussions concerning reactor mishaps and media politics, the conversations became a safe retreat from the day's prime focus.

What were the symposium speakers presenting? Primarily the reality that over 40,000 nuclear weapons and their delivery systems now existed, and that they were at times brandished in such a hair-trigger fashion that they might be unleashed even by minor accidents. The weapons' victims would not be some demonic "enemies" but ordinary humans, both in our own country and in other nations. Without citizen action, sooner or later the bombs would be dropped. I resisted this message by allowing my own career demands, my own sociability and my own passive optimism to blur its impact. I did this partly because I had other priorities and partly because I was preoccupied, but largely because I did not wish to think about nuclear war.

Even the generation that had most strongly contested American military policy—my own—in a strange way assumed that atomic warheads were manufactured to sit and rust. We didn't want them increased, of course. We didn't support their production. We surely had little faith they made us safer. But they existed, and always had as we grew up during what we termed, rather fatalistically, "the nuclear age"; we'd been immersed in their presence long enough to disregard them.

Dr. Helen Caldicott closed the conference by comparing nuclear danger to the disease of a terminally ill patient. We would have to face the usual immobilizing reactions of denial, anger and grief; but if we gave our effort all the care and human commitment possible, we just might be able to save life on the planet. As I was leaving, thinking the conference had eroded a bit of my fatalism, a psychologist handed me a survey on attitudes toward nuclear weapons.

I resented it initially; statistical polls seemed part of the same fetish for abstraction that had brought us to this clear and present danger to begin with. But whether or not the questions worked as research, they made me think. When I reached the section on whether I had ever talked with my family about nuclear war, I admitted we'd never discussed the subject.

But there is no appropriate time, no sanctioned context, no accepted relationship in which we are supposed to think or talk or act with the goal of confronting the atomic threat. Society has banned the topic from thought or discussion with a compulsion nearly as powerful as the force of the nuclear bombs themselves—and my acquiescence to this silence was precisely what the symposium sought to challenge.

No one, the speakers said, had a greater or lesser responsibility to challenge weapons that made all humans potential targets. Arms questions could no longer be left to scientists, journalists or activists, to politicians, think tank specialists, or least of all, to the generals who'd commissioned the weapons in the first place. To create a public dialogue on what we'd allowed to develop in our name, we had first to confront our own denial that the mechanisms to end human civilization existed.

Except for a few phenomena such as the Cerenkov effect,

nothing in the Hanford plants' physical presence makes them particularly fearsome. To be sure, the old reactors convey a certain Promethean menace as they rise out of the desert like demon machines from *Metropolis*. But images of hissing steam valves, dark corridors and smokestacks making the sky glow with their toxic effluents are all from an era which atomic technology was designed to replace. The colors here are not the reds and yellows of primeval furnaces or the black of the raw iron they produced, but the white, silver and hospital green of any controlled and sanitized environment. Whereas Vietnam opposition had an impressionistic backdrop of tracers, body counts and jungle rot, citizens confronting the dangers of Hanford must settle for images of deer grazing by the electrical towers, clouds piling high and red as in a William Blake painting while the sun sets on the Columbia, and wind blowing across sagebrush plains so calm and immense that it's easy to believe no human actions could ever affect them.

How long have people lived in a situation in which we encounter overriding dangers that are both delayed and invisible? The depredations of carnivorous beasts, of storms, of floods and droughts were all concrete and immediate. But now people are confronted with threats that cannot be seen, that have effects surfacing years or even generations later—and, in the case of nuclear warheads, that have not directly taken human life in thirty-five years.

Though Hanford's exceptional environment has furthered these risks, most of its residents treat it as routinely as did the Kennewick high school girl whose mother came home from Hanford saying "Guess what? My shoe got crapped up and they had to burn it." "But it was no biggie," the girl explained. "She went upstairs, took a shower and never mentioned it again." I asked if she herself had ever worried or talked with classmates about possible dangers. She lit a cigarette, played with her gold neck chain and answered: "No. Do the older people?"

Many trivialize, like the Hanford wives who compare spills

of radioactive waste liquids to "flat tires that tell you when it's time to change something," who rationalize leak coverups as "mopping up your own kitchen mess instead of telling all the neighbors," and who consider Three Mile Island "a good thing because you don't stub your toe twice."

Most just go in every day to the Area because America is a country of empiricists, and radiation effects do not show up in bank accounts, do not appear on screens, and cannot be touched, seen or heard. They go in because they are pioneers, and "if Lewis and Clark had let minor problems stop them we'd never have migrated west of the Mississippi." They go in because nuclear work is simply the job that they do.

11
Rebels and Dissenters

Ray Goldsmith—an Area manager in his mid-forties who as a graduate student had edited a published anthology of bawdy folksongs, and who, although he "wasn't quite an SDSer," had supported his students in their Vietnam protests—was a bit apprehensive when he came to Hanford six years before from the East Coast. He had worked with poisons and carcinogens, and had even bid on a contract for a nerve gas warning system. But he was skittish around radioactivity and got kidded for it.

Those worries—specific to his own safety—passed quickly, though, and as we talked in his new Richland home while Charlie Parker riffs drifted up from the basement where his son was practicing the saxophone, Ray addressed nuclear dangers in philosophical terms. "The rest of the world should at least have a chance at the best aspects of our standard of living," he said. "That doesn't mean everyone has to have three cars and a dozen TV sets. But it also doesn't mean going back to nothing but wood stoves—we don't even have enough trees to support them anyway—or to the coal that, when I was growing up in Brooklyn, turned every sheet we ever hung outside a dirty gray.

"It means keeping and disseminating to other cultures what's worthwhile in our medicine, our agriculture, and in the technologies which allow us to amply clothe and shelter our inhabitants. If we want these things, we're going to need either conservation and recycling or tapping off-planet sources, but we're also going to need almost unlimited power. The breeder of fusion research might yield that. Even though there are uncertainties, those benefits would be worth whatever potential risks might be encountered."

Countering arguments about the hazard of nuclear wastes, Ray cited cities where lead—which is toxic and never decays—is already concentrated over limits set by the Environmental Protection Agency and contended that, from a technical viewpoint, the environmental impact of covering a region with coal plants or solar collectors isn't necessarily less than that of nuclear power. "Oil," he said, "is used for so many chemical processes, it's a crime to waste it as an energy source. And while Nagasaki is now reinhabited, we have defoliated portions of Vietnam which may never be."

As a scientist and a creator, Ray remembered with pride developing a synthetic pheromone which—by confusing male gypsy moths "till they ended up balling the plants instead of each other"—helped save a major portion of Pennsylvania's agricultural crops. When the first vat was processed, his team went in at four o'clock in the morning and signed their name to the container; he'd felt the same excitement producing pesticides "which put more food on people's tables all over the world."

Useful technologies, he explained, leaning back beneath a large Mondrian-style painting, were adaptable; he hoped whatever he helped produce would be fitted to the social contexts it was used in. Ray felt like a teacher here, "watching hesitant young Ph.D.'s become mature scientists," and he linked Hanford's research and energy production to world-scale needs: "Speaking pragmatically, questions like the potential for international famine can't be evaded. Our ability to survive as a human species is related to the resources we can develop and use. I hope," he hesitated for a moment, "that my kids have as much personal freedom as I do. I'd be terrified if the need for nuclear security brought on the government keeping national records on us all. But I also hope they'll be able to at least cool the house to 80 degrees in the summer and heat it to 65 in the winter; that they'll be able still to travel and to see the country;

that they'll have energy available when and how they need it . . ."

In this town where people keep doubts to themselves or toss them off in casual bitch-session asides, Ray's twenty-nine-year-old friend Steve Stalos was the exception who had made his criticisms public—by resigning his Rockwell job in charge of monitoring one of the waste storage "tank farms," by running for Congress against pronuclear incumbent Mike McCormack and by bringing to media attention declassified but buried information, including reports on radioactive gas leaks and on plutonium contamination found in coyote feces and in two dead mice discovered in a Hanford vending machine.

When I met Stalos in a coffee shop near Richland's Uptown Shopping Center, his beard, rimless glasses and easygoing presence reminded me of a graduate student in an Athens café. Explaining that he liked his co-workers here, Stalos emphasized that they weren't malign; "They just lack critical perspective." He pointed to a glass of water on our table and said it contained radioactive ruthenium and tritium from the slow seepage of wastes into the Columbia. Although the concentration probably wasn't enough to hurt us, the isotopes shouldn't have been there; most Hanford workers didn't even know they were, and releases of wastes into the environment always created the potential for hazard.

"I'm no neo-Luddite," Steve told me with an edge of frustration. "I think much new technology—like the lasers which can reattach severed retinas—clearly improves human life, and if they could handle everything right I'd actually support nuclear energy. But all accidents—whether they're caused by faulty welds, poor operation or mistakes in site and design choices—involve some kind of 'human error.' Even though Rockwell would hardly go broke doing a decent job on the tank farm monitoring here, it would be prohibitively expensive to create the checks and backup systems you need all along the length of the nuclear cycle. If you count in the government's handling of insurance, waste disposal and plant decommissioning, plus the need—so we won't run out of uranium in twenty-five years—for unproven and highly dangerous breeder technology, nuclear power isn't cheap at all."

Although Ray had told me that he thought Stalos would still support nuclear energy for Third World countries lacking other sources of power, Steve said building reactors would leave those nations with "not only the normal risk problems, but also a dependency on highly complex technologies which they can afford even less than us . . . This whole project began," he continued, "not because it made sense, but because two brilliant men—Fermi and Bohr—figured out it was possible. You could build and draw power from a mile-high drinking duck [the child's toy which bobs up and down powered by evaporation of liquid from its beak], but that doesn't mean it would be wise to go ahead and do it."

Stalos came to Hanford at the beginning of 1974, after asking his Montana State physics professor if he thought nuclear power was a good growing field. Assigned to ARCO's Personnel Protection Group, he initially found the work interesting and worthwhile. But he began to wonder when, sent to work with radiation monitors after a few months here, he realized they were confusing the basic terms used in measuring damage to personnel by radiation. "The monitors would say 'He received 50 millirems,' a statement impossible to make by reading the dials of their instruments, because the roentgens per hour that they measured were completely different units than REMs." After going through all his books trying to figure things out, Stalos stopped believing the crusty old monitors when they insisted repeatedly that they knew what they were doing. When he finally went to the higher-ups they explained, "Yes, the monitors do have it wrong. But retraining would be confusing, and besides, everyone knows what they mean."

The saga of the Hanford waste tanks began in the Area's earliest days when, given the challenges of building and running the initial reactors and separations plants, disposal of what would become the world's largest quantity of radioactive by-products was viewed as a detail to be worked out later. As an interim solution, Du Pont built steel tanks 75 feet in diameter, installed them underground and used them to store radioactive liquids and sludges left over from the plutonium extraction process and general operations. More were built to accompany the General Electric era reactors. When Stalos became manager for tank farm surveillance analysis in April 1977, there were 149 of them, plus 7 newer ones with double-shelled walls.

Each tank contains from 55,000 to one million gallons of wastes, some caustic, some several hundred degrees in temperature, and all emitting radioactivity to further stress steel that would rust out sooner or later even if it contained only water. When tanks leaked—and between 1958 and 1975 twenty were announced as having failed—the cesium, strontium and plutonium contained in the waste solutions percolated down through the soil below until eventually at least a fraction of the material reached the water table.

Because the tanks are underground, monitoring them for leaks is complex. Traditionally, surveillance teams lowered meal probes through pipes (some of the workers also used the pipes to lower ashtrays on strings so the gamma rays would turn them into amber souvenirs). When the probes hit the liquid surface they sent back an electrical charge. If a tape had to be lowered farther on a particular test than on the one preceding, a liquid loss had probably occurred.

As long as the tanks were full, this method was fairly accurate—and liquid level monitoring uncovered a number of major leaks (including one, in 1973, of 115,000 gallons) that have released a total of 450,000 gallons of high-level wastes into the Hanford soil. But since an evaporation program has reduced the volume of tank-held material from over 100 million gallons in 1972 to around 40 million at present, the liquid level measurements have become a far less effective means of finding leaks. Much of the solute wastes are now mixed in with salt cakes at the bottom of the tanks. The salt cakes are irregular and give varying readings depending on where on their surface the probes are lowered. Even where tanks have been only partially pumped, floating crystalline sludges make consistent measurement impossible. Because the pumped tanks still contain thousands of gallons of presently irretrievable liquids mixed in an atomic gumbo with the salt cakes, and because even the solid material will eventually eat through the walls and leak into the soil, the evaporation program has by no means solved all problems.

In recent years surveillance teams have gained their most useful data from dry well readings. Workers, often wearing protective suits, lower instruments through pipes placed in the ground surrounding the tanks and assess any surges above background radioactivity. If such increases occur (the soil unfortunately shields radioactive presences more than six feet from the instruments), the teams locate the direction of the highest readings and drill additional wells to trace the origin of radioactive material.

Stalos managed the group responsible for analyzing leak detection data, and it was less than a month after he took the job that his team discovered their first abnormal readings. Noting what his senior technician called "a slow increase in activity" between tanks 103TX and 107TX, he ordered the drilling of additional test wells to determine the source. The first, placed to intercept a hypothetical leak from tank 103TX, showed nothing. But, with the aid of a highly effective detection instrument borrowed from Battelle, the team traced the direction toward the north—toward 107TX—and confirmed their suspicion when another well, three feet from the tank's side wall, produced readings over one hundred times as high. Stalos's supervisor, John Deichman, agreed that the tank should be pumped immediately. Within a few days 90 percent of its contents were removed.

Three months elapsed between the initial reading and the draining of the tank. In the meantime, the radioactive plume had traveled to two more in-line wells, and geologists had ruled out its being the result of horizontal soil migrations from earlier spills. Stalos wrote up his Occurrence Report, labeling 1O7TX a confirmed leaker, and presented it to Deichman for his signature. But Deichman, as Stalos recalled in testimony connected to a subsequent Department of Energy investigation, said a DOE superior had told him not to report 107TX as a leaking tank. "Steve," he said, "I promise we'll call this a leaker. Just give me some time to soften up the boys downtown."

Stalos was surprised at this comment and at Deichman's reluctance to report the leak. He asked if there was any doubt they'd correctly identified the source of the leaked material. Deichman said he had no doubts, that the other suspected tank would be returned to active service, but that the report filed would be inconclusive and interim.

A month later DOE Senior Staff Engineer Chet Compton asked Steve to meet with him and suggested—as an alternative possibility to a leak—that the wastes might have overflowed the tank through poorly plugged pipes, or that (based on a different geological interpretation) the radioactive materials might well have been deposited in the soil from earlier spills. As Stalos was leaving, Compton admonished him, "Think about what you're doing, Steve. A man gets a reputation, and it can hurt him his whole career."

During the next six weeks, until January 5, 1978, "interim" reports continued to be filed on 107TX, each stating that the investigation was continuing and that the source of the radioactive material had not been determined. When Stalos pressed for more information, he kept hearing that the matter was being taken care of, that he wasn't making friends for himself in the company by pushing it, and that the DOE people could have a lot to say about his future advancement.

After eight interim reports, Deichman, together with Rockwell's Waste Management Program Director, issued a final Occurrence Report approved by company General Manager Donald Cockeram. It stated that the cause of the radiation increase had not been determined, that the "application of additional investigative resources" did not appear "cost beneficial," and that there was no indication "that a specific apparent cause could be identified." But since the soundness of Tank 107TX could not be "positively substantiated," it was classified as of questionable integrity and restricted from further use.

During and after the 107TX controversy, Stalos and his surveillance team came up with dry well readings indicating leakage in two other tanks. Again there were only highly speculative alternative explanations for the data. And, as before, the investigations were no longer deemed cost effective after a certain point and the confirmed leaker status of the tanks was denied.

Sixteen tanks were classified as of questionable integrity during the period from the beginning of 1976 through the end of 1979—and in some cases the surrounding ground was so saturated with radioactive material that acurate judgments concerning the origin of any radioactive material were, in fact, impossible. But while numerous tanks were investigated, none were judged to be leaking.

In July of 1978, Stalos received a copy of a letter from Alex Fremling, manager of the Department of Energy's Richland field office, to Rockwell head Cockeram, which suggested dry well monitoring be cut back. On November 8, Stalos was ordered in writing to reduce previous weekly readings to a biweekly schedule. Though he protested that resulting delays in leak discovery might mean over a thousand gallons needlessly spilled in any given incident, he received only the now standard explanations regarding cost constraints and manpower limits. Stalos resigned on December 5, 1978, left for a nonnuclear project at Maryland's Godard Space Flight Center six months later, and testified on Hanford waste management practices at a Senate subcommittee hearing held at the end of 1979.

When I met him, Rockwell head Don Cockeram turned out to be a genial, professorial man who, until he moved to the outskirts of Kennewick so his wife could run a kennel, rode his bicycle 30 miles to and from the Area each day. Because of the waste evaporation, he explained, it was no longer possible to tell which specific tanks had leaked. True, the test wells indicated radioactive presences. "But that's like finding drops of water beneath a sink and saying a particular pipe has rusted through; if you call one a leaker you're saying it isn't any of the surrounding ones instead, and doing that would be being delinquent in our responsibility."

"But weren't some clearly leaking?" I asked.

"More than likely, but Steve wants us to say we're positive, and while maybe we should call them 'possible leakers' or something more specific, it just wouldn't be right to say we were sure."

When I asked about monitoring cutbacks, Deichman explained that they were already pumping the liquid out of all the old tanks anyway, and since surveillance was "just information," wasn't it more important to put all their efforts "into doing"?

The talk ended as Cockeram leaned close to me, dropped his voice as if to confide a special state secret, explained "of course they will all fail eventually"—then said it would be nice "to confirm some specific leaks just to help our general credibility," but he couldn't in all conscience target tanks he wasn't sure of.

I asked Deichman later about his confrontations with Stalos, and he explained that pumping the 107TX tank was consistent with questionable integrity status. He denied saying he'd "soften up the boys downtown" and denied ever indicating the tank should be confirmed as a leaker. "Because I didn't agree," Deichman said, "I stepped out of the way to give Steve his day in court before an executive management board. We didn't have both the dry well and liquid level evidence needed to call it a definite leaker. That's why the board decided against him."

Chet Compton also insisted Stalos had misconstrued his words—saying he cautioned Steve about his reputation and future career, not because he supported any cover-up, "but because I was saying in effect, 'You're a sharp young man. You've got a good background. Don't just run off saying things without checking all the facts.' " Again, Chet mentioned piping overflows as a possible explanation for the dry well increases and suggested the radioactive material might also have been deposited in the early days, "when if they had a spill within the tank area they often didn't even bother to write a report."

Because the waste tank controversies involved differing interpretations of phenomena such as dry well readings and soil movement patterns, as well as differing accounts of conversations, it was hard to know who to believe. But Stalos was essentially vindicated by the DOE Inspector General's report on "Alleged Cover-ups of Leaks of Radioactive Materials at Hanford."

The document began by summarizing Steve's charge that the classification of questionable integrity was used so Hanford management could deal with leaks by pumping the probable source tanks, filing highly technical Occurrence Reports, and avoiding the public announcements that might bring negative attention to the industry. Because, the report said, confirmed leaker classification required supportive evidence in all but the most extreme cases from both dry well and liquid level readings, and because levels in pumped tanks could no longer be accurately measured, "The solidification program has . . . made it much harder, and in some cases, impossible, to acquire the evidence that must be gathered for a tank to be classified as a confirmed leaker."

While the investigators made no final judgment as to whether Hanford management created a conscious cover-up, they suggested that Rockwell reexamine criteria used for tank classification, reexamine the classification categories themselves and review "policies relating to when public announcements about tank leaks will be made."

"The word 'cover-up,' " the report concluded, "evokes pictures of people devising strategies and tactics aimed at concealing things which ought not to be concealed. But in the case of Hanford, had there been any officials desiring to minimize publicity about tank leaks, they would have had no real need to engage in conduct which might be considered questionable. This is because Hanford's existing waste management policies and practices have themselves sufficed to keep publicity about possible leaks to a minimum."

Given the assaults on atomic power by citizens' organizations and individual critics, Hanford management is less than enthusiastic about helping potential opponents pick over the industry's sores. While both the Senate hearings and the DOE investigation were going on, Rockwell did finally reclassify four questionable integrity tanks as confirmed leakers. All had derived their status around ten years earlier from liquid level readings, and after lengthy controversy the company determined that no dry-well evidence was needed. But, as the cover-up study explained, a group of Rockwell middle managers also met and decided their report on the tank status changes should not be volunteered to the investigation. When the investigators, informed of the report by an anonymous source, asked the Rockwell officials about this meeting, the managers said at first that they had no recollection "of any such discussion or decision." Then one admitted his powers of recall had often been faulty and suggested they believe their source.

But this was only putting one's best foot forward for visitors—which was why, Rockwell Hanford operations chief Dale Bartholomew explained to reporters from the *Seattle Post-Intelligencer*, he removed markers indicating the boundaries of a radioactive contamination spread the night before a team from the *Washington Post* came through for a story in 1978. Another time Bartholomew scrawled a warning, on a document assessing the spread of windblown radioactive particles, that if Rockwell higher-ups formally accepted the report, antinuclear groups could use the Freedom of Information Act to obtain its damaging information. But that was, he said, just concern that the report would be "misconstrued."

Difficult as Hanford's guardedness made things—and I encountered so many stalls, bureaucratic evasions and misleading answers that I often thought I was playing a top security game of tag—the situation is still better than in England, where the Official Secrets Act prohibits employees of major atomic facilities from discussing their work at all, and official censors decide whether or not one can write about a variety of nuclear industry aspects.

But Hanford needed neither an Official Secrets Act nor, as the Investigator's Report said, official cover-ups. Though some old hands were less sanguine about atomic risks than others, the attitudes of most here resembled those of a Hanford manager who responded to a *Post-Intelligencer* report that swallows were building radioactive nests with mud from the ditch where the laundry handling the contaminated clothing dumped its waste liquids by saying "that water is so safe I could drink it for years and not be harmed." The waste sites, the veterans argued, both in official statements and internal conversations, are 25 miles away from Richland and 150 feet above the water table. The basalt formations below are nearly impermeable. And if radioactive material, in addition to the present short-lived isotopes, does reach the Columbia, it's really nothing to worry about.

Leaving aside the very real question of nuclear economics, one could explain, as Stalos did, that stricter safety precautions would mean more, not less, Area jobs—and that in waste management, at least, there'd be a need "for Hanford or its equivalent as long as there are humans on the North American continent." But the old hands want to put their energy into building and doing, not checking, doublechecking, filling out forms and becoming the equivalent of nuclear janitors. With Stalos's concept of responsibility bringing their every action under minute scrutiny, its general application would make it impossible for the men to get the job done as they always had.

Nearly everyone I spoke with had heard of the controversy over the waste tanks. And because Hanford had never before produced public dissenters, and because disposal of wastes remains one of the nuclear industry's most critical problems, Stalos—like a West Point man rooting for the Viet Cong—was regarded as a fifth column threat to the commitment of all who worked here. If this project should fail, I was constantly reminded, it would be due not to technical problems but to politics, and now this onetime insider was handing out the fortress keys to all potential enemies who asked.

If you didn't know Stalos, it made things easier. You could joke at Christmas parties about establishing "Stalos Memorial Leaker Awards," or assume, when he resigned, that he'd actually been fired. You could label him, as did Bob, the hovercraft builder, "a pipsqueak son of a bitch who shouldn't have been in surveillance because he was antinuclear from the start." You could decide he was on the take, as did former WPPSS board president John Goldsbury, who—after telling me that he'd met an antinuclear woman dumb enough to believe his statement that when the Hanford lights went out the workers' ears started glowing—said, "I heard Stalos was always getting in trouble for coming in drunk and failing to show up on time, and that if he hadn't quit they would have terminated him anyway. Now he has it a lot easier. He's getting paid $3,000 for his speaking tours, and I know this because he told me." You could explain, as did Sam Beerman, "Now I don't know Steve, but I can't imagine Don Cockeram doing anything unethical."

If you were involved in the controversy, it was more complicated. You could treat Steve simply as a natural-born malcontent, as did John Deichman, who explained when asked about Stalos's motivations, "Well, he ran for Congress on an antinuclear platform. He made a big public deal about resigning. All that's a matter of record." Or you could react like Chet Compton, who said he guessed he felt kind of paternal toward Steve. "He was going off half-cocked like all those people—Gofman and Tamplin and Nader—who are at odds with the scientific community. He didn't realize that even the earlier dumping of wastes directly into the ground didn't really harm anything. It seemed like he was sort of a far-out dreamer, running against Congressman Mike McCormack as a dark horse, looking for a cause to get involved in and happening to find the nuclear one. Steve thought his dry-well readings came from the 107TX tank and wouldn't discuss other options, but it's hard—when you've been around twenty years and he's only been working two or three—to understand what he was so upset about.

"It's like those three engineers who worked for GE and left a few years back," Compton continued. "They'd all come to the system relatively recently. They only knew a little about it. From what they understood there was cause for concern, but instead of checking out the whole picture they kind of grabbed what they saw and ran with it."

That Stalos's judgment could have been reasonable was inconceivable, but because Chet was a kind man, he attributed Steve's dissent to youth and to ignorance, just as he did that of the three GE engineers—who actually averaged almost twenty years of nuclear industry experience apiece.

"Steve was friendly, intelligent and a lot of fun," said a young technician who'd worked under him. "We used to play catch at lunch time, and whenever I had a problem I'd enjoy going over to him and talking it over. We even had something in common, because my wife grew up in the same Montana town where his uncle owned a bowling alley, and sometimes we'd talk about that . . . I respect him for holding his views and sticking to them."

"What did you think of the issues he raised?" I asked.

"Well, anyone leaving the company, the management tends to give bad reports. Steve had his side. They had theirs. Obviously, if I disliked him I wouldn't still send him a Christmas card, but I was just a technician logging data and it's not my place to judge these things."

I asked about the reactions of other workers, but he answered only that where he grew up (on a farm 80 miles away) he was taught "that talk didn't tell anyone anything . . . and I guess

I'm just the kind of person who doesn't hold opinions on other people." Whether on or off the record, he refused to comment on the tank controversy, instead smiling a lot, apologizing and repeating again and again that he just didn't think it was his business to judge.

Though the rupture Stalos caused by going public was unprecedented, and though he left expecting that he would not be the last Hanford dissident, I wondered how his departure would affect reactions to future safety problems. Before, people could hesitate and think, "Well, maybe this Stalos guy's charges do make sense." They could feed information to Steve, and he could pass it on to the press. Now, lacking the support and catalyzing spark of a publicly critical presence, they'd have to take the risk of becoming whistle-blowers on their own.

Although he arrived in the Tri-Cities shortly before Stalos left, Creg Darby attempted to facilitate these links between outside dissent and the day-to-day workers. Darby's "Hanford Conversion Project" originated when Northwest antinuclear activists decided that, since Hanford affected the entire region, a special group to confront its issues would be useful.

The organizers secured grant money to fund subsistence salaries for Darby and for another worker in Portland. Creg arrived with a brief list of potentially sympathetic names, and with his hopes both to serve as a watchdog and to foster a Tri-Cities challenge to the all-powerful overlord, the Area. When I met him in August of the same year, he had written several pamphlets on nuclear hazards, had gone on speaking tours of Yakima, Walla Walla and other nearby towns and had put together a small antinuclear demonstration when Nader came to debate the nuclear proponents. Creg held press conferences when the DOE decided to start up the FFTF for a test run without first installing a safety mechanism recommended by a Nuclear Regulatory Commission committee. He kept a log of nuclear problems that were buried in obscure stories in the *Tri-City Herald*.

But Creg's hopes of organizing were never fulfilled. Although friendly engineers invited him over for spaghetti dinners, xeroxed pronuclear articles for him and told him stories of their days in the atomic Navy, the milieu ground Darby down. His tiny Pasco apartment, with its bare light bulb, cracked linoleum and Salvation Army furniture, could have fit into a single room of one of the boomer's lavish spreads. Although, as a former physics major, Creg knew his issues and was a good researcher and writer, he seemed overwhelmed by Hanford's immense scale and pace, culturally isolated by the predominant complacency and suffocating under the poverty he was living in. Deciding that changes in Hanford could come only "from Americans elsewhere saying 'Sorry, you may like this, but we don't and we won't fund it anymore,'" Creg left a month after I met him. Six months later, the project's Portland office closed as well.

Battelle Hanford's solar energy program straddles a line between challenging the Area's basic assumptions and serving as a deferential complement. It developed, ironically, out of the nuclear industry after twenty-year atomic veteran Kirk Drumheller's kids came home from their liberal western Washington colleges and began questioning him about alternative energy possibilities. A Hanford colleague informed Kirk that 3,000 megawatts fell on each square mile of earth on a sunny day. When Drumheller checked out available literature on the subject, the potential of that 3,000 megawatt "magic number" began to spark his interest. Though he still worked half-time studying options for nuclear waste disposal, he got his first Department of Energy solar grant in 1971; by 1980, Battelle had one hundred people studying how to derive energy from the sun, the wind, biomass, geothermal concentrations and low-head hydroelectric projects, plus fifty others researching conservation.

As we met in his Battelle office, Drumheller described the virtues of soft technology as the practical empiricist he continued to be. He gave me books full of fuel cost/energy output data tables. He explained how new developments in semi-conductors, plastics and lightweight metal alloys made many of the renewable technologies more feasible. He charted on a blackboard the distinctions between end-uses for which almost any form of energy would be sufficient, and those specifically requiring high-grade sources like electricity. He cited the difference in payback time between winterizing, designing new, energy-efficient houses or installing solar hot water systems—all of which used existing technologies and required at most a half dozen months to begin yielding benefits—and constructing the massive oil, coal or nuclear plants that required years. He explained that although many here still saw nuclear power as the key to a healthy industrial future if one used the Nuclear Regulatory Commission's maximum estimate of 166 reactors on line by the year 2000, they would supply at most 10 percent of America's energy needs; that using the conservative figures compiled by Robert Stobaugh and Daniel Yergin's Harvard Business School Energy Project, conservation by the same year could save 30 to 40 percent and solar technologies generate between 7 and 23 percent of what we needed; that many of these renewable technologies were, even at present, less costly than the nuclear option.*

Drumheller believed most people here felt solar energy could not be a major contributor to electricity production, industrial process heat or transportation, but said recent years had created a change in some attitudes. Because the necessary engineering skills transferred easily, many of his colleagues were ex-nuclear people "who'd still go back to atomics if this program folded," and who in some cases even spent their mornings on reactors and their afternoons on solar projects. Tri-Cities now had a half dozen solar-related businesses and an active solar energy organization, at a recent meeting of which

*According to Drumheller's figures, nuclear electricity costs an average equivalent of $2 per gallon compared to $1.54 per gallon for large, mass-produced wind systems, $1 to $1.58 a gallon for solar hot water heating and 62¢ a gallon for passive solar design on new buildings.

Kirk had counted nearly one hundred people (and at which one man got a fair number of signatures on an initiative petition to place spending limits on WPPSS). Some of the tinkerers had even installed energy-efficient heat pumps in their homes and designed their own solar hot-water systems or swimming pool heaters. Because the Battelle hands were more pragmatic doers than philosophers, Drumheller was still almost the only one who read such theoreticians as Amory Lovins, Barry Commoner or Denis Hayes. But both Kirk and those he worked with hoped ultimately to develop devices like low-cost photovoltaic cells that would turn solar power directly into electricity and allow unlimited consumption "if we could once get the technics down."

Because Drumheller works at Hanford, for the DOE and Battelle, some solar people mistrust him, and he's been denied speaking engagements after making clear his equivocal but not entirely critical position regarding nuclear energy. His Tri-Cities solar speeches include obligatory statements of how he would, of course, rather live next to a nuclear reactor than a coal plant. He is reluctant to suggest that, leaving safety considerations aside, it might well be projects like the $1 billion FFTF or the $24 billion WPPSS plants that deny renewable resources the capital they need to become our energy future.

"I tend to be impatient," Kirk said, when I asked how others here view his projects, "with people who don't appreciate solar's potential right away. I have to keep reminding myself that I too worried about its generating ability and thought of it at most as a long-term alternative. It wasn't until I'd read everything I could for years that I became a kind of solar nut. I shouldn't expect people in other fields to understand the technology instantly.

"One of the most basic problems in the energy business is that there simply isn't time for everybody to learn about all the possible alternatives. Someone working a 60-hour week in the nuclear business doesn't have time to keep up on what's going on in solar; someone working similarly in solar can't keep up with nuclear developments." People therefore developed understood maxims—like the insistence that solar couldn't possibly make a contribution for twenty years—repeated them to their colleagues and set it up so the maxims were in turn used to respond to the questions of others farther down the line. "Finally you ask the one guy who does know, he gives a different answer and you don't believe him."

The result was that Drumheller would show friends documents explaining both passive and active energy systems in terms of materials required, land area used and BTUs produced; two weeks later they'd tell him, "You're just giving me this because you're pro-solar." The same people who wouldn't hesitate to invest $5 or $6 million of company money in a nuclear system Kirk recommended would listen to his discussions of heliostats—the rotating mirrors which reflect light onto generating surfaces—and say only "But you have no idea how much they'll cost." They'd tell him, "Well, I guess solar's not much good today," each time the sky was cloudy, and when he'd explain that wasn't exactly true, would switch to one of the other instant objections that they swallowed as casually as morning corn flakes. Drumheller spent years trying to refute "the core assumptions that machines using the sun or wind are somehow hoaxes or un-American plots."

The old hands I spoke with regarded Kirk not exactly as a traitor, but as a pie-in-the-sky dreamer allied vaguely with the nefarious no-growthers. Skepticism was expressed affectionately by Clark Reitnauer, Drumheller's former partner in a short-lived although reasonably successful goggle manufacturing business. "I just don't agree with him," Clark said. "You'd need hundreds of acres of solar collectors and it just wouldn't be economical or practical. Of course," he laughed, "I didn't think jet planes would fly either. If people really think it will work, they should put their money where their mouths are, invest in the industry and make a million dollars."

More often, though, the old hands would say "Kirk's a nice guy, but . . ." and go on to label his studies "just a bunch of numbers" with a vehemence more usually reserved for such direct critics as John Gofman and Ernest Sternglass. Drumheller himself didn't think we should quit working on other systems but admitted, when pushed repeatedly, that limited resources should go first to the renewable technologies.

Hanford has had a few other dissenters: the varying "moles" who, while they never revealed their names, became so angered at sloppy procedures that they leaked information on buried incidents so that outside reporters, particularly those from the *Post-Intelligencer*, could pressure Hanford's public relations staff into giving them corroborating explanations and documents.

But people who criticized in public here were politicians—"damn politicians" in the phrase power-broker Volpentest used when Stalos got up and asked questions at an NRC briefing—or else outcasts like those Henrietta Beerman envisioned when she inquired whether I was "one of those Three Mile Islanders." Never mind the government subsidies supporting every project and every worker here. Never mind the giant nuclear corporations and the petty hustles that facilitated advancement within their ranks. Hanfordites were rugged, individual loners; dissidents were ragged, scraggly group-think victims. Hanfordites walked tall, with families in tow; dissidents were "bandwagon types looking for causes." The outsiders huddled together in their marginality; Hanfordites continued onward in their task of building.

Unfortunately for those opposing technologies which may be progressing close to the tipping point in their irreversible effects, Hanford and the other "nuclear parks" that seem to be the industry's most likely future are harder to challenge than complexes located in normal communities. To confront the technology, critics should understand both concrete risks such as those of reactor accidents and leaking waste tanks and the less tangible threats of a culture in which institutional decisions are never questioned. And they should be aware as well of how these dangers manifest themselves in day-to-day aspects of the nuclear cycle and in the lives of those employed at the reactors. In part, learning this is contingent upon putting in sufficient time, energy and discipline. But the information must be available. For the Hanford people themselves, legacies of the original security regulations, like the censorship of letters by Military Intelligence and the bans on discussing work with outsiders, limit access both to this information and to the viewpoints that might offer the workers critical perspectives on the institution they serve. And because informal prohibitions succeeded legal codes when the official restrictions ended, a circle of silence makes it easy not to question.

Laying aside the myths that Hanford, like Topsy, "just grewed," it still appears to those here as a finished, immutable institution. If you quit in protest you'll be replaced. If you dislike things badly enough you can leave as Stalos did, as Darby did, as Kendall Ozaroff did, or as did all the others who decided Atomic City could not be their home. If you try to confront the mission or how it's carried out, you risk not only being branded a pariah, but also being overwhelmed by the slow crush of bureaucratic resistance. So in this place where nuclear dangers are most intimately encountered, almost no discussion deals with their implications.

Most here diffuse challenge both by immersing themselves in their work and by trusting the integrity of their friends and neighbors. Although there are people here who are fearful, mistrusting and, in Woody Guthrie's phrase, "just plain mean," they exist no more so than in any other community. And when Tri-Cities residents think of men such as Sam Beerman, John Rector, Clark Reitnauer or Lester Dumont, how can they decide an industry these workers helped create could in any fashion be slipshod or immoral?

Innocence—the desire to ascribe all tragedies to the fates—makes sense in a context where the effects of one's actions are limited. But untarnished refuges are an illusion when mundane routines of going to work can contribute to the deaths of humans thousands of miles away. They are an illusion even when one does not work directly at Hanford or some equivalent facility, but merely supports the weapons operations and the commercial atomic projects by paying taxes, by participating in a highly interconnected economy and by failing to challenge a political climate which allows the manufacture of thermonuclear warheads to be viewed as a normal and respectable job. They are an illusion when innocence becomes coupled with the death of social consciousness, the death of critical questioning and the death of citizen participation in the most urgent decisions of our time.

It does not seem overly harsh to say the men and women who work on atomic weapons distance themselves from the moral implications of what they do. Because the menace of the atom's commercial cycle lies not in the product's designed use but in side effects nuclear proponents claim are preventable, the ethical responsibility of workers involved in nonmilitary nuclear projects is more complex. But in either case, the atomic technologies' unprecedented possible consequences should demand more than passivity and silence from those who work with them.

The nuclear enterprise is obviously not the only potentially destructive one to have proliferated in a society where inventors and builders are shielded from the judgments of both ordinary citizens and of isolated intellectual critics. But as security and geography separate complexes like Hanford from those who challenge their mission, and as a new class of bureaucrats emerges to smooth over all problems, the same isolation that allows nuclear workers to proceed unquestioning in their tasks also makes it far easier for the rest of us—who ultimately pay the social, economic and environmental costs, and who would become fellow casualties were the atomic weapons ever used—to safely distance ourselves from what it is we are supporting.

Can the proliferation of both the "peaceful" and warlike atoms be reversed? Primitive technologies—such as those of wood, water and animal power which preceded the fossil fuel explosions—have often been superceded by more complex, presumably more efficient systems. But when have entire technological matrices been abandoned, not just because they were "obsolete" or "undoable," but because the risks they entailed were unacceptable?

Within the atomic industry itself, examples exist of paths begun and later pulled back from: a nuclear-powered airplane which would have been able to run forever without refueling but whose shielding requirements would have made it impossibly heavy; the "Orion Project" of physicist Ted Taylor, which, until bans on atmospheric blasts made further testing impossible, planned to explode atomic blasts against a massive shield to propel a gigantic vehicle to the stars; the notion, given up in the design stages for fear of accident consequences, of using rockets to dispose of nuclear wastes by shooting them into the sun. But these were fragments rather than rejections of wholesale means of taming the world.

While at the Physicians for Social Responsibility symposium, I asked *Bulletin of the Atomic Scientists* editor in chief Bernard Feld whether precedent existed for pulling back from enterprises developed on such a massive scale. Feld said the only partial comparison he could think of was with the post-World War I scientists who'd united in refusing to work on biological weapons and with the action of the League of Nations in successfully banning their use. I thought of two non-Western examples: the Japanese, who gave up possession of guns after deciding their use was un-warriorlike, and the Ming Dynasty Chinese, who, after sending expeditions involving thousands of people to India, Africa and other far-flung places, turned their backs on overseas colonization and burned their ships and their navigation maps to prevent future efforts. But those were vastly different cultures. The nuclear order is strongly established and well entrenched. Feld thought it would take a massive, though essential, effort to reverse the trend.

Ironically, just as the Reagan era pragmatists were suggesting atomic war could be a reasonable tool with which to further American interests, the atom's commercial side suffered a substantial defeat with the construction halt of WPPSS plant 4 at Hanford and 5 at Satsop.

The decision had its immediate roots in WPPSS's massively overextended borrowing. Because the Supply System's tax-free bonds were over half those sold in the United States, because they were increasingly saturating the market and because the borrowing seemed endless and the energy payback continually postponed, a number of brokerages commissioned major studies on the security of this particular investment that they were promoting. Their reports, written between 1980 and 1981 by a group of young securities analysts at Merrill Lynch Pierce Fenner & Smith, at Drexel Burnham & Lambert and at Wertheim & Co., doubted whether WPPSS could absorb the ever-increasing debt to which they were committing themselves. They doubted whether the Northwest should build plants whose cost canceled out the advantages of cheap power that had attracted industries to the region in the first place. And they doubted whether the additional power, whose sale WPPSS assumed would eventually repay all costs, would even be needed at all.

For those in charge of WPPSS construction, problems had, of course, been building long before the critical reports were released. Contracts would be readjusted, completion dates pushed forward and another $1, $2 or $3 billion added to the estimated budgets. The WPPSS heads would line up another $150 to $200 million worth of bond buyers. The interest rates at which investors could be convinced to participate would keep rising—from 6 percent on the bonds sold in 1978 and 1979 to between 10 and 12 percent on bonds sold in late 1980 (because the bonds were tax-free, they still went for far lower rates than did conventional utility offerings). Although WPPSS managers debated accepting the debt load, no other sources existed for the money.

Then in May 1981, just nine months after Robert Ferguson

had been made the new WPPSS head, a budget estimate came in at $24 billion—$4 billion higher than anyone had remotely suggested was possible. According to the new figures, WPPSS would have to borrow $3 billion in each of the coming three years. A large percentage of this would finance plant 4 at Hanford and plant 5 at Satsop. The resulting interest load threatened the financial solvency of all five projects.

The Drexel Burnham and the Wertheim reports were already released by this point, and the conclusions of Merrill Lynch were also known. On May 26, 1981, Ferguson suggested that a one-year moratorium on plants 4 and 5 would allow WPPSS to complete the other three and solve its financial crisis.

Just as the WPPSS debt was unprecedented, so there was no parallel to the notion of two multi-billion dollar reactors left to sit because of lack of money. But the funds for full-scale construction could not be reasonably obtained. The WPPSS debt load was already staggering for the ratepayers of Washington and of participating utilities in nearby states. After the WPPSS board held several weeks of desperate conferences (including one session in Oregon to avoid Washington's open meeting law), they accepted Ferguson's recommendations and three months later agreed to a two-and-a-half-year cessation of construction.

Ferguson initially attempted a half measure—a slowdown, he explained, not a total halt. But the bonds for the two plants had now dropped two rating points to classification BAA (the lower the ratings, the less secure the investment and the higher the interest rate). Raising even $1 billion instead of the original $3 billion seemed unlikely. WPPSS gave up, dropped its work force at WNP 4 from 3,400 to 50 and at WNP 5 from 700 to 50, and—as the moratorium threatened to be a permanent one—began pushing the Bonneville Power Association for a federal bailout.

For the WPPSS construction workers, the halt was less catastrophic than would seem likely. True, site work stopped, cranes left, half-constructed buildings evoked the feel of an instant ghost town. But aside from some migrants, most craftspeople were quickly absorbed into the labor forces of the other two plants.

Lou Hansen's swing-shift crew of eighty welders, inspectors and supervisors reported one night as usual to WNP 4, then were told as they were getting ready to leave that they were to show up at WNP 1 the next night. "The adjustment wasn't difficult," Lou said, "because the plants are identical, and a hole in the floor of 4 is in the exact same location as 1. You don't get lost. You don't get confused. You just pick up where you left off and do your job." At first workers talked a lot, he said, "about which of WPPSS's Machiavellian motives caused the decision." But after a while even that ceased, and they simply performed their tasks as usual, taking the same "whatever happens happens" approach they used in so many other aspects of their lives, resting their hopes for ongoing employment in the endless building of plants 1 and 2, assuming more atomic wealth would arrive in the form of other projects in the future.

One can view the WPPSS crash in nonnuclear terms: call it a runaway boondoggle finally brought to ground, and explain how mismanagement led the plants to cost far more than other similar-sized ones built at the same time. One can praise at least the possibility of marketplace justice for ratepayers whose still immense debt load may now be slightly reined in. One can say the reactors fell by their own weight (though they may well be resurrected by federal intervention) and decide that objective economic laws will decide the atomic industry's fate. One can trust these laws and a supposed free market to guarantee that reactors will only be built and operated if they in fact represent the greatest good for the greatest number.

But in a sense, those who say the environmentalists shot them down may be correct. Were it not for the public outcries, these reactors and all others could have been built, as the Du Pont ones were, without complicating inspection requirements, redundant (though certainly still vulnerable) safety systems and design changes when plants elsewhere revealed new and previously unsuspected flaws. The industry could have taken the approach of an atomic Navy man, quoted approvingly in the *American Nuclear Society Magazine*, who explained to a colleague in the commercial sector that he should do as they did: "Just put the sum bitches in and don't say anything about them." Confronting infinitely fewer checks and limits, the plants would probably have been completed before the costs of WPPSS's immense mismanagement soared to levels that made continued work as usual economically impossible.

The closing of atomic projects for fiscal reasons can no more be separated from citizen resistance to the industry than the end of the Vietnam war could be from the dissenting movements that forced the removal of American troops and prevented a seriously considered use of atomic weapons.

If one views plants such as 4 and 5 as examples of worthy technology capriciously hampered, then demonstrations and challenges clearly represent an obstructionist effort. But if, on the other hand, nuclear power portends not a necessary future but an unwise danger that at the very least must be held strictly accountable for all its consequences, then the shutting down of the two WPPSS reactors for what were in the immediate sense financial rationales was still a victory for human reason and human community.

12

Waiting for the Angels

One day Rick Price, the Hanford engineer's son who went off to study British history, showed me a 1950 government booklet entitled *Survival Under Atomic Attack.* "You can live through an atom bomb raid," it reassured its readers, "and you won't need a Geiger counter, protective clothing or special training in order to do it." Instructions detailed how wearing a hat with a brim and loose-fitting light-colored clothing could save you from flash burns; how your toilet tank was a good source of uncontaminated water; how your chances of recovering from radiation poisoning were "much the same as from everyday accidents . . ."

"Just like a match-book ad," Rick said. " 'You too can win great prizes or become a star radio announcer.' It would be great if fallout were no worse than a burn from a kitchen stove or a spill off a ladder. Now I admit I thought this booklet was pretty strange when I found it at a friend's house when I was seventeen. But I heard the same crap the whole time I was growing up and they still feed a slightly more realistic version to anyone who comes in all bright-eyed to work at the Area. They told us repeatedly that while Hanford was special the work didn't place our dads in any danger. And we knew that even if we'd created the power to destroy the world, we were tough enough to handle it."

These sentiments were repeated, with a celebratory edge, in another item of old Hanford memorabilia: a copy of the *Sage Sentinel* newspaper which circulated here around the wartime construction camp. In an issue printed shortly after the Japanese surrender, an "Interview with the Atom" was sandwiched in for comic relief between tips on pie-making and photos from a recent Hometown Week party.

"I get it," began the interviewer's question. "The Japs felt your splitting fit of laughter in Hiroshima. But what were you doing in Nagasaki?"

"Seeing if it's true," the atom replied, "that as it says in the song, 'Back in Nagasaki, the sailors chew tobaccy and the women wicky-wacky-woo.' "

"And do they?"

"Not any more. . . ."

Hanford today remains sufficiently secure and guileless that ten-speed bikes are left unattended on Richland front lawns, that people I'd just met would invite me into their homes and fix me tunafish sandwiches, that the home phones of Sam Volpentest, of the Rockwell head and of the local Department of Energy chief are all listed.

But two-thirds of the Area workers I interviewed spoke only because their names were to be changed or comments kept off the record. Their fears escalated when the moratorium on WNP 4 and 5 raised the specter that America just might withdraw its subsidies and leave the atomic enterprise to crumble. And many here, I became aware, had convinced themselves that the world in general was about to end.

I first noticed this three days into my initial visit here. After watching the Creative Anachronists' sword tournament, I asked N reactor safety specialist Chuck Matthews how he liked them.

"They were pretty accurate," he said. "But I've studied medieval history and I noticed three of their characters mixed, into the same costume, clothing from Tudor England and from fifteenth-century France."

After explaining that what I thought was the picture of a butterfly on my shirt was actually that of a luna moth, he asked if I'd read Thomas Malthus. "He's not real popular now, but

I'm a 100 percent Malthusian, and I've studied the historical boom and famine cycles which prove he's right. Rome, you see, declined from a population of over 1 million to 20,000. The French Revolution was caused by overcrowding. Moorish culture only flourished centuries after the North African city of Carthage was destroyed." As Chuck explained that in order to have a golden age a culture had to collapse and then grow again, I looked at the soaring eagle on his belt buckle and at the black glasses, crew cut and rumpled blue shirt which made him resemble Disney's Absent-Minded Professor. I saw the anguish on his face and in his gestures as he strained to communicate his revelations. He cut from Constantinople plagues to crises in the bond market and the ravages of inflation. Comparing American culture to "a society running down the street being chased by a bulldozer," he hoped that the breeder could create unlimited fission power before time ran out. Whatever peace we've ever had, he said, was due to our huge space, plentiful resources and small population; but with all frontiers now closed, necessary growth required exactly that energy we were about to run out of.

"I've studied these things," he said once more. "I've drawn up charts and figures. Of course I support conservation 100 percent, but without the power of thousands of plants like the ones we're building, political unrest will make our country uninhabitable anyway.

"You know, Hitler was a Malthusian. Now don't misunderstand me, I'm not saying what he did was good, but given his context, if he hadn't created those terrible slave camps they would have had starvation, revolution and civil war. Yes, it was a dictatorship. No, I'm not a Nazi. But people just won't admit he had no alternatives. And they won't admit now that our choices are less than we think."

Maybe, as pastor Joe Harding explained to me later, Matthews was an atypical loner. Maybe, as Stalos suggested, he was "one of those whom that long brooding bus ride just gets to after a while." But I kept entering presumably normal conversations and hearing people such as Mark, the Pasco farmhand, and Cindy, the young Exxon worker, tell me, "Well, you know the Book of Revelations says it's all going to end real soon." There were the official apocalyptic believers as well, such as Duane, the Pentecostal kid, and those who played their pessimism straight and cynical, among them Steve Girea with his comparisons of the human race to "a company going on past its time." Some here—such as a twenty-two-year-old Kennewick flower child into incense, old Joni Mitchell albums and chocolate ice cream—were more naive, talking about how the atoms would "get pissed off and make this place the end of the world." Some even brought in quite rational disaster specters, as did Ray Goldsmith in an unsettling and quite convincing discussion of world famine possibilities; Stalos mentioning the likelihood of nuclear war; or a young, environmentally concerned Kennewick railroad worker who worried about what would happen if a major Hanford accident destroyed the Columbia River watershed.

Perhaps the expression of all these sentiments was coincidental. Perhaps I was wrong to perceive them as being voiced more often here than in any other place I'd ever been. And the dour old hands did seem to express them far less frequently than did the young and more displaced generation which succeeded them. Because this was just a place where people worked their jobs and drew their paychecks, because Nagasaki was thirty-five years back and because tarring nuclear power's commercial end with holocaust associations was perhaps to confuse a common genesis with a common future, I resisted seeing a preoccupation with apocalypse. Still, in the same manner in which people have passed on attitudes regarding security, mission and protective silence, perhaps some special fear still lingered in the Hanford air from the initial quantum leap in the power of humans to destroy.

On January 3, 1961, a small reactor at the government's Idaho Falls test site blew up. One of its graveyard shift operators had been manually lifting the main control rod the four inches necessary to re-link it with the regular gear mechanism after disconnection for maintenance. When he yanked it too high the reactor went supercritical, and the resulting steam explosion blew off the pressure lid. The operator precipitating the accident was impaled on the ceiling by a broken control rod piece which shot instantly through his groin. The effects of the blast killed his two shiftmates. The bodies were so contaminated they had to bury one of them in a lead-lined coffin.

No one was sure why the operator erred at a task he'd performed so many times before. But eighteen years later, just before I made my initial Hanford trip, it was discovered that the man pulling the rod had apparently done it to gain revenge on one of his shiftmates, who'd been sleeping with his wife. The explanation reminded me of Z plant, and of a young operator at the FFTF who worried about going rock climbing with two co-workers, again, involved with the same woman, because she wasn't enthusiastic about "hanging below one on a rope the other might be holding." It reminded me also of one of Boyd Norton's stories from his *Audubon* piece:

It was 1968, and Norton, working as a supervisor at another Idaho Falls reactor, left the control room to a young physicist (he called him "Arnie") who'd been growing increasingly apprehensive about an appeal of his draft status. Because they were starting the reactor for a test, Norton was supposed to be directly overseeing; but busy in conversation about some of their data, he told Arnie to go ahead without him. It turned out Arnie had just been discussing, with another employee, the likelihood of his being sent to Vietnam. He was shaky and spacey as a result. He pulled the rods too fast. Lights and buzzers went off. A fellow operator scrammed the reactor barely in time.

Thinking about these incidents and about the possibilities of some millennial dreamer attempting to turn this small corner of Washington State into the fire-and-brimstone Armageddon that appeared to be coming anyway, I wondered what it meant for believers in apocalyptic destruction to be working here—

and whether nuclear workers should receive special leaves of absence for times when their emotions were upset, or when their heads just ached. Tri-Cities leaders had already tried (and failed) to combat drug use with their anti-paraphernalia law; perhaps the logical next step would be a return of the old security days—with tighter checks, of course, to correspond to the increased temptations of a more hedonistic time.

But eliminating all the dopers might leave Hanford without a construction force. There was enough resentment over bureaucratic oversight without adding psychiatrists with white note pads to complicate the picture. And, as participants in a recently held Menninger Foundation Stress Seminar reminded an audience of Hanford managers, it was never possible to completely screen out workers with severe psychological problems.

When I first met psychologist Mike Casey, he told me the story of a local VIP ("I can't tell you who he was, but you'd probably recognize his name") who came up to him in a supermarket and said, "Mike, I think in ten years the angels are going to come."

"I sort of looked at him," Casey recalled—"I mean, if he had been a client that would have certified him psychotic—and mumbled 'Uh-huh.' 'Well you're a Casey, you're a Catholic,' he said. 'You understand that, don't you?' And when I answered that I was a Northern Ireland Casey and that I didn't understand, he disappeared."

A year later I talked with Casey again. He told me now that the VIP was a Tri-Cities notable unconnected with Hanford and that he'd cited him merely to indicate that crazies existed at all levels and in all positions. The effects of psychological stress did concern him, though, because you couldn't check and doublecheck all activity, and he worried about the worker "who goes in for his shift in the middle of a divorce suit, or after his kid has run away in the night or who drinks too much or takes drugs." But he didn't think this place was either more or less apocalyptic than any other.

I wasn't sure. I wondered whether—the precise details of the angels story aside—Casey had changed his mind during the intervening period. Because how could one really say that the product Hanford workers created led them to private fears different from those experienced by people anywhere else? I recalled a conversation I'd had with Virginia and Lester's son Robert. Now a microbiologist in Seattle, Robert described growing up with plenty of kids his own age to tramp around with in the desert, watching Lester assemble a cherry wood cabinet worthy of any master craftsman, going to high school and splitting his time between the academic crowd and the kids who preferred spending lazy afternoons playing softball. If his childhood was somewhat sheltered, it was also, on the whole, quite happy. But Robert also knew three women on his block and the next one over who periodically broke down, began screaming unceasingly and had to be hospitalized. He mentioned the Harris boys, two sons of a Hanford engineer, who ended up killing themselves in consecutive years. He remembered another wholesome Richland kid who died in his parents' house from a heroin overdose. These tragedies obviously happened everywhere, and, to the degree that they were rooted in this particular environment, they were no doubt caused as much by Hanford's isolation and transience as by its product. But Robert thought the work and the craziness just might be connected.

"I'm hesitant to draw it out," he explained. "There were some bizarre family situations, to be sure. The words 'death wish' come to mind, and I wouldn't dispute the possibility or even probability of some link. But I can't completely say this came from working on the ultimate weapon. Banal as it sounds, I suppose people's reaction to their situation was mostly to become blasé."

Perhaps whether atoms did or didn't breed nightmares really didn't matter. If contributing to weapons which could bring the angels for us all did in fact terrify some here into either religiously sanctified or private secular desperation, could workers in such states of crisis perform, with perfectly focused attention and intellect, each of the minute and varied tasks of the nuclear cycle? If, on the other hand, most people here felt not fear but a sense that both their own persons and the nuclear systems in general were invulnerable, and that any weapons products would never be used, could they then avoid the carelessness and complacency bred by 40-hour-a-week, 48-week-a-year jobs anywhere?

Several weeks into my first visit to Hanford and thirty-six years after the residents of the original towns were evicted, I attended one of the annual August reunions during which the onetime pioneers drive into the Area and tour their former homesteads. Now in their fifties, sixties and seventies, the women and men caravanned out to the old town of White Bluffs, then stood talking in small groups with old friends or acquaintances from previous years. They watched the deer and quail that had succeeded them as inhabitants here. They looked out at the old orchards once blooming full with their efforts but now stripped bare. Their houses were gone, of course, leveled in the Manhattan Project days so as not to provide cover for spies or saboteurs. The sole remaining structure—Hanford High School—loomed out of the desert, its white walls and faded green lintels sagging like Miss Havisham's wedding cake, left to crumble after a ceremony that never took place.

The old-timers entered the building tentatively, lowering their voices as if talking too loudly might cause even their memories to disappear. Inside, broken pipes headed nowhere and iron rods stuck out, like lost insect feelers, from the concrete they once reinforced. Partitions separated classrooms, halls and gymnasiums, then ended abruptly in broken floorboards and jagged rubble. Empty window frames looked blankly out at the orchards, the sagebrush and the sand.

Because the old trees, despite their eerie starkness, remained lined in their original neatly planted rows, they brought back images of frame houses and white picket fences, of quilting bees, freshly squeezed lemonade and back porch swings—of the small-town Norman Rockwell America which the suburbanites of Richland and other bedroom communities sought to re-create in the midst of modern affluent convenience. But most of these old-timers lost their ties to that world when they lost their homes. They came now from Seattle, Salt Lake City and Baltimore. They were (or had been before they retired) not farmers but electricians, secretaries and teachers. They passed their time talking of their grandchildren and remembering incidents like the time when the Columbia froze solid and they had to lead a herd of sheep across through wagon ruts they made in the ice.

Although some had sued for more just government compensation following the war, most accepted the necessity of their dispossession. Now, with the weight of life in other places overshadowing what once was here, discontent surfaced only in passing references to parents who wished they could be buried where they grew up, or in talk of those whose finances flourished elsewhere, but who "would have given everything up" for the tiny Hanford homesteads the government gave them ten days' notice on. When I asked about the evictions, Helen, a store clerk in her mid-fifties whose shoulder pin displayed an electron whirl surrounding the words UP-N-ATOM, answered, "At least they didn't put other people in our place. But if Du Pont had spent all that money on irrigation, we'd have such a beautiful garden country now. Instead they put that whatchacallit—that atom in it, and now it's ruined forever. I guess that's progress but I don't know."

Helen spoke of the displacement as if the "progress" it represented was something she could not challenge. The reactors had been built and the old towns leveled. The fruit trees were stripped far more thoroughly than they'd ever been by the "damn Okies" her friend Roy remembered as always stealing his apples. Like the residents of "that there Hiroshimmy," which another woman thought "maybe also got a bad deal," no one prepared these people to be caught up in nuclear destiny. Maybe their lives were just expendable war-effort resources to be used interchangeably like money, tanks or trucks. But because what these onetime pioneers had built was destroyed for other uses, they returned this day only as tourists visiting their own past.

A few here mistrusted the legacy of those first-in-the-world production reactors: Two argued about waste storage, the Columbia being ruined and one woman's fear that her grandchildren "would never even know what it was we grew up with." But for most, Hanford's present endeavors existed only as backdrops to the past they'd lost.

As we drove farther into the Area, I felt the echo of a desert wind which could blow from stillness to 60 miles per hour in minutes, with a dry beating heat like that of Southern California's Santa Ana or the Arabian hamsin. The air was quiet now as we passed the first of the earlier reactors. Their stepped roofs and gray smokestacks made them resemble nineteenth-century coal plants or New Deal-era public housing tenements rather than anyone's dream of glistening white futures. They seemed at a quick glance more archaic than apocalyptic.

After leaving the main reunion party, I accepted the invitation of a Portland fruit and vegetable salesman named Bob Codding to join his family in a ride out to visit the place where he grew up.

The Hanford Grange Hall held Depression-era dances where, for twenty-five cents Bob used to hear fiddlers, drink as much beer as he wanted and "go round and round until I was ready to drop on the floor." But the hall was gone. All that remained was a musty, cobweb-filled bank complete with a useless safe and dangling light fixtures, an empty brick power station guarding a half-ruined spillway that descended into the Columbia and a cracked sidewalk running from the river to another ghostly orchard 60 yards in. Black locust trees, plus a few hardy poplars, survived nearby without need of human tending. On the opposite bank, by the striated clay escarpments which had once served as a meeting place in Chief Joseph's 1877 Nez Percé revolt and which later had named the White Bluffs town, a construction worker ran checks on a high-tension tower and played Bruce Springsteen on a portable radio.

Bob's house was, of course, gone from its former site. Instead, grass grew in the old bulldozer ruts and a Model T Ford lay with its doors on the ground. Nearby, on the site of the old town warehouse, a pile of detritus made me imagine Simon Rodia coming here to glue and cement this motley assemblage of poultry incubators, empty boot-black tins and broken pots into the Hanford equivalent of Watts Towers. But when I looked out at the flanking H and D plants which now sat squat and dead a quarter mile away, I realized this place already had monuments which would remain on this plain many times longer than human history; for even if the concrete reactor structures were "decommissioned" to a further stage and taken apart to be buried in nameless sites, the ground they would leave behind would be saturated for over a hundred thousand years with radioactive by-products from the weapons-grade plutonium.

I was about to leave the site when I spotted an old rusted ax blade. Though no longer usable—the iron loop which once provided a socket for its handle was crushed—I envisioned the ax being swung to hue and construct a world which existed before the atom came here.

When I looked up, the trees between where I stood and the reactors were lined, as before, in neatly ordered rows. Their trunks cracked and twisted as they reached out for the final drops of water which humans had first brought, making their growth possible where only grasses had existed before, and then removed. With the scent of long-dead wood hanging on the dry air, the legacy of man-made objects forever untouchable began to bring back post-holocaust images from old science fiction paperbacks. As a hoot owl called from a dark fallen tree, Bob walked slowly, with his hair graying and back hunched, through the cheat grass and stands of wild rye.

"We used to gather driftwood from the river and lie on the hill here," he explained, "and hide and play ball in the gully below . . . They bulldozed all of it flat for 'security.' Now I can't even find the remains."

I grasped the ax—a mute token of the past—and heard Bob's daughter joke with her boyfriend about kissing being illegal on the reservation. Bob had stopped walking now, remembering perhaps the time "when jackrabbits ran so thick they almost covered the alfalfa," but more likely wondering how this place where he was born and raised had come to be erased like this. Thinking maybe it could at least serve as a keepsake to trigger memories, I showed him the ax and asked if he wanted it. "No. That's all right," he answered quietly, and thanked me. I looked at it again and—though I hesitated to play the role of souvenir-grabbing tourist—asked the young patrolman who'd joined us whether he saw any problem with my taking it. When he told me not to worry and explained that he collected old bottles here all the time, I placed the ax in my car and headed back to town and to a pot-luck picnic I wanted to attend.

But when I visited Stalos later in the evening and mentioned bringing back the ax, he did a doubletake: "Where did you go? What did you touch? You used this glass here, didn't you? The chances are maybe one in ten thousand that it's crapped up; but you have no way of knowing that some leak twenty years ago didn't deposit radioactive material there, and if it did, you've just endangered every person you shook hands with or shared food with since you drove back."

Perhaps because it was Steve Stalos calling to see where the ax could be checked out, a Rockwell manager was on the telephone in minutes, explaining that, although they didn't have a radiation monitoring station in Richland (and the situation didn't seem critical enough to send a mobile team), they'd be happy to take care of things at the nearby 300 Area. When we got there, after a ten-minute drive, the first alpha particle detector didn't work. After the young woman operator banged on it a few times unsuccessfully, she decided "there must be a short somewhere," and came back after a brief search with another machine. The reading was negative for both the ax and my car seat where I'd left it; they lectured me, as if I were a ten-year-old, on how everything in the Area is government property; my fingers, shortly afterward, stopped tingling nervously.

But I thought about hot zone workers who, after repeated checks found them "clean," decided precautions were "company bullshit" and stopped following them; the Z plant kids, and how the burning plutonium caught them so entirely by surprise; some guards fired the previous month for taking discarded metal home to make belt buckles. I thought about the casual creation of the raw material for H-bombs, and the eventual surrender of pride in craft and responsibility for product. When I came originally I had viewed this place primarily as a major site for an energy option I mistrusted because of physical occurrences such as waste leaks and mechanical failures—and I would have been nervous just entering the Hanford Area. Now, like everyone here, I had begun to treat the environment as routine.

Afterword

It is in many ways fitting to end the Hanford story with the potential links between complacency and annihilation. This controlled and dominated environment does breed a culture of political passivity. And sentimental optimism will not ameliorate the consequences of what is produced here.

Yet it is a trap to use the nuclear culture's lessons to validate despair. However scarce were the men and women who, like Szilard and like Stalos, broke from the atomic institutions to challenge them, their critiques catalyzed the resistance of others. Though it was and still is far easier to leave Tri-Cities than to stay and create a public dialogue, Americans as a group can no longer evade what they have created by constantly moving on to open frontiers. And when the public sufficiently challenges the atomic enterprise, it turns out—as in the case of WPPSS plants 4 and 5, and the aborted reactor in Skagit Valley—to be something other than an unstoppable monolith.

If one accepts the assumptions of pragmatism, what exists becomes inevitable; speculations on futures outside the given political, social and economic arrangements become utopian and useless. In a sense, immersion in the overwhelming presence of the Area and its culture reinforced my own pessimism. Insofar as I resigned myself to recording a culture bent on doom, I ceded power to affect a future which I, together with other ordinary people, would inhabit. I began to settle for understanding the rules of the WPPSS board game.

But if Hanford is a microcosm of a society in many ways out of control, it is also a community. And even the members of this circumscribed environment possess character strengths that might hold possibilities of hope. True, these strengths are checked by what Robert Lifton calls "psychic numbing." but could the desires of the old hands for quality work and sense of purpose serve causes more humane than the manufacture of thermonuclear weapons, and could the young atomic boomers use their critical imagination to envision a culture they could embrace without cynicism? Could friendships developed through 30 years of bridge clubs, church services, and Kiwanis luncheons begin to speak for more than affluent "security"? Could the modest familial dreams be prototypes for a society that lives within its means and values trust and reciprocity more than acquisition?

Whether a political dialogue and political alternatives can address latent values and latent needs like these will determine whether or not nuclear culture becomes an inevitable human future. And that is a task not to be left to "the men who know best," but to be undertaken, as best we can, by all of us.

Acknowledgments

For all that I am critical of the Hanford mission, most Area workers, on a personal level, went to substantial trouble to help me understand the project in which they'd vested their lives. Because the towns they live in are small ones where occupational and social sanctions are easily imposed, I have changed the names of most of the people in the book to protect them against potential repercussions; therefore thanking specific people here is impossible. But to all in Tri-Cities who gave generously of their time, their memories and their perceptions, and who allowed a critical outsider into their intimate lives, I hope you will view this book as an honest effort to grapple with the nature of your community.

Although this is the first major published work on Hanford, I found Paul John Deutschmann's 1952 University of Oregon Ph.D. thesis, "Federal City: A Study of the Administration of Richland, Washington, An Atomic Energy Commission Community," quite helpful in understanding Richland's development as a company town, and Ted Van Arsdale's booklet, *Hanford, the Big Secret*, useful for the information it contained on the war years. In terms of secondary sources on nuclear issues, Robert Jungk's *Brighter Than a Thousand Suns* remains, twenty-five years after its initial publication, essential for anyone wishing to understand how we entered the nuclear age. Jungk's recently published *The New Tyranny* provides a frightening look at the relationships between nuclear technologies and authoritarian social systems. For a basic understanding of reactors and their operation, I found Walter C. Patterson's *Nuclear Power* extremely useful, as I did Amory Lovins's *Soft*

Energy Paths and Barry Commoner's *The Politics of Energy* for analyses of nuclear economics and alternatives to atomic power. Two Three Mile Island studies—Boyd Norton's May 1980 *Audubon* article entitled "Supercritical" and Daniel Ford's essay in the April 6th and April 13th 1981 issues of the *New Yorker*—gave detailed explanations of how nuclear accidents actually occur. For discussion of atomic war and of how we distance ourselves from its implications, I learned much from Robert Jay Lifton's books *History and Human Survival* and *Death In Life*, and from a March 1978 article by Ron Rosenbaum in *Harper's* entitled "The Subterranean World of the Bomb." Lastly, Helen Caldicott's *Nuclear Madness* provided a passionate moral argument for citizen action opposing nuclear weapons and power.

Although my sources for understanding Hanford's cultural aspects are far more diverse than those for understanding its technology, I found a number of studies of instrumental rationality and of bureaucratic distancing particularly helpful. They include Lewis Mumford's *Technics & Civilization*, Hannah Arendt's *Eichmann In Jerusalem*, Henri Lefebvre's *Everyday Life in the Modern World*, and Theodor Adorno and Max Horkheimer's *Dialectic of Enlightenment*.

On a more personal note, it was the encouragement of numerous friends that kept me writing from when I started, eight years ago, to the point where I was ready to do this book. Some of these people were intimates, some simply co-workers on common projects, but all helped keep me hopeful and keep me going. Thank you Pete Knutson, Ron Williams, Stanley Aronowitz, Laura Knoepler, Rich Stites, Carol Prentice, Jean Moshofsky, Michael Cohen, Benita Hack, Lorna Schiede Milgram, Lorraine Pedersen, Stewart Weiner, Warren Allen, Kathy Shagass, Alan Bernstein, Bruce Trigg, Gwenda Blair, Gary Drieblatt, Sebastian Milito and Mark Powelson.

A number of people reviewed this manuscript in whole or in part, and improved it with comments on varying sections. They include Ed Dobb, Gary Soucie, Magda Loeb (who has also been a very good mother), Steve Stalos, William Appleman Williams, Joel Connelly (who helped enormously in explaining the tangled maze of WPPSS and whose articles, co-written with Eric Nalder in the *Seattle Post-Intelligencer*, initially interested me in the subject) and Dan Chasan.

To conclude, there are four people without whom *Nuclear Culture* would not exist in its present form. Tony Astrachan took a chance as a *Geo* editor on a young, at the time only slightly published writer, and gave me the magazine assignment that allowed me to go out and make my initial research visit to Hanford. Liz Gjelten vested painstaking and enormously skilled effort in helping me tame, organize and edit the tangled mass of characters, impressions, histories and implications that ultimately became this book. Joanne Dehoney not only came up with continually useful suggestions at all stages of the manuscript, but was a friend and more whose humor and caring had a lot to do with keeping me sane through an endless parade of 70-hour work weeks. Finally, James Monaco encouraged me from the time when he first suggested I attempt to publish a paper I had written for a class he taught. Thank you, Jim, for insisting that I keep trying.